www.EffortlessMath.com

... So Much More Online!

- ✓ FREE ALEKS Math lessons
- ✓ More ALEKS Math learning books!
- ✓ ALEKS Mathematics Worksheets
- ✓ Online ALEKS Math Course

Need a PDF version of this book?

Please visit www.EffortlessMath.com

10 Full Length ALEKS Math Practice Tests

The Practice You Need to Ace the ALEKS Math Test

By

Reza Nazari

All inquiries should be addressed to:

info@EffortlessMath.com

www.EffortlessMath.com

ISBN: 978-1-63719-462-1

Published by: **Effortless Math Education Inc.**

For Online Math Practice Visit www.EffortlessMath.com

Welcome to

ALEKS Math Prep

2023

Thank you for choosing Effortless Math for your ALEKS Math test preparation and congratulations on making the decision to take the ALEKS test! It's a remarkable move you are taking, one that shouldn't be diminished in any capacity. That's why you need to use every tool possible to ensure you succeed on the test with the highest possible score, and this extensive practice book is one such tool.

If math has never been a strong subject for you, don't worry! This book will help you prepare for (and even ACE) the ALEKS Math Assessment. As test day draws nearer, effective preparation becomes increasingly more important. Thankfully, you have this comprehensive practice book to help you get ready for the test. With this book, you can feel confident that you will be more than ready for the ALEKS Math test when the time comes.

First and foremost, it is important to note that this book is a practice book and not a prep book. Every test of this "self-guided math practice book" was carefully developed to ensure that you are making the most effective use of your time while preparing for the test. This up-to-date guide reflects the 2023 test guidelines and will put you on the right track to hone your math skills, overcome exam anxiety, and boost your confidence, so that you do your best to succeed on the ALEKS Math test.

This practice book will:

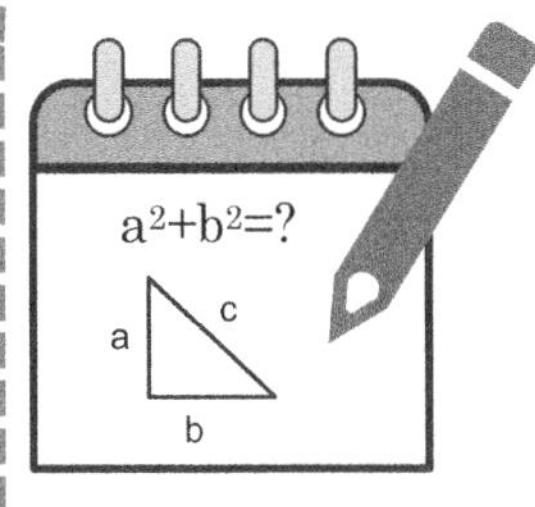

- ☑ Explain the format of the ALEKS Math test.
- ☑ Describe specific test-taking strategies that you can use on the test.
- ☑ Provide ALEKS Math test-taking tips.
- ☑ Help you identify the areas in which you need to concentrate your study time.
- ☑ Offer ALEKS Math questions and explanations to help you develop the basic math skills.
- ☑ Give **realistic and full-length practice tests** (featuring new question types) with detailed answers to help you measure your exam readiness and build confidence.

This practice book contains 10 practice tests to help you succeed on the ALEKS Math test. You'll get in-depth instructions on every math topic as well as tips and techniques on how to answer each question type. You'll also get plenty of practice questions to boost your test-taking confidence.

In addition, in the following pages you'll find:

- **How to Use This Book Effectively** – This section provides you with step-by-step instructions on how to get the most out of this comprehensive practice book.
- **How to study for the ALEKS Math Test** – A six-step study program has been developed to help you make the best use of this book and prepare for your ALEKS Math test. Here you'll find tips and strategies to guide your study program and help you understand ALEKS Math and how to ace the test.

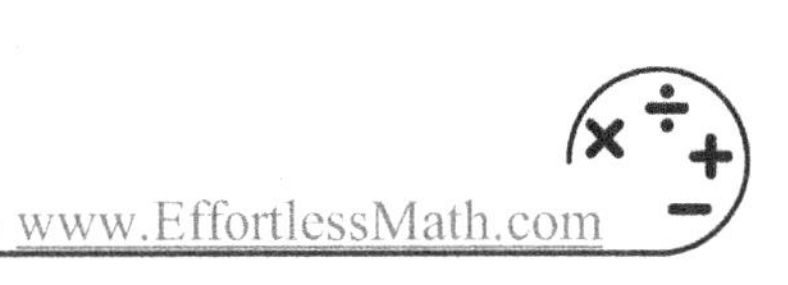

- **ALEKS Math Review** – Learn everything you need to know about the ALEKS Math test.
- **ALEKS Math Test-Taking Strategies** – Learn how to effectively put these recommended test-taking techniques into use for improving your ALEKS Math score.
- **Test Day Tips** – Review these tips to make sure you will do your best when the big day comes.

Effortless Math's ALEKS Online Center

Effortless Math Online ALEKS Center offers a complete study program, including the following:

- ✓ Step-by-step instructions on how to prepare for the ALEKS Math test
- ✓ Numerous ALEKS Math worksheets to help you measure your math skills
- ✓ Complete list of ALEKS Math formulas
- ✓ Video lessons for all ALEKS Math topics
- ✓ Full-length ALEKS Math practice tests
- ✓ And much more...

No Registration Required.

Visit **EffortlessMath.com/ALEKS** to find your online ALEKS Math resources.

How to Use This Book Effectively

Look no further when you need a practice book to improve your math skills to succeed on the math portion of the ALEKS test. Each section of this comprehensive practice book will provide you with the knowledge, tools, and understanding needed to succeed on the test.

It's imperative that you understand each practice question before moving onto another one, as that's the way to guarantee your success. Each practice test provides you with a step-by-step guide of every question to better understand the content that will be on the test. To get the best possible results from this book:

- **Begin studying long before your test date**. This provides you ample time to learn the different math concepts. The earlier you begin studying for the test, the sharper your skills will be. Do not procrastinate! Provide yourself with plenty of time to learn the concepts and feel comfortable that you understand them when your test date arrives.

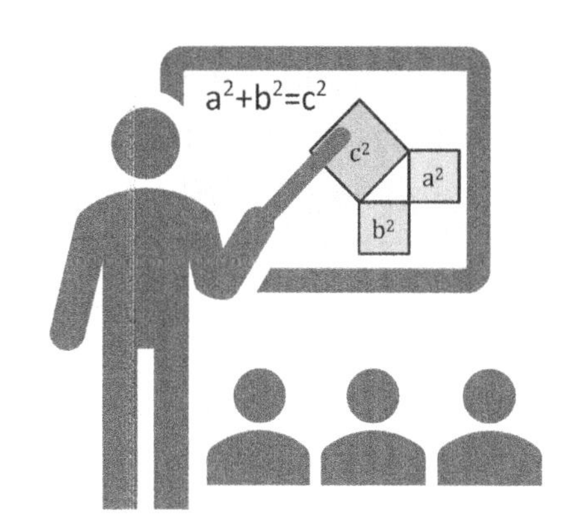

- **Practice consistently**. Study ALEKS Math concepts at least 20 to 30 minutes a day. Remember, slow and steady wins the race, which can be applied to preparing for the ALEKS Math test. Instead of cramming to tackle everything at once, be patient and learn the math topics in short bursts.
- Whenever you get a math problem wrong, **mark it off, and review it later** to make sure you understand the concept.
- Once you've reviewed the book's instructions, **take a practice test** to gauge your level of readiness. Then, review your results. Read detailed answers and solutions for each question you missed.
- **Take another practice test** to get an idea of how ready you are to take the actual exam. Taking the practice tests will give you the confidence you need on test day. Simulate the ALEKS testing environment by sitting in a quiet room free from distraction. Make sure to clock yourself with a timer.

How to Study for the ALEKS Math Test

Studying for the ALEKS Math test can be a really daunting and boring task. What's the best way to go about it? Is there a certain study method that works better than others? Well, studying for the ALEKS Math can be done effectively. The following six-step program has been designed to make preparing for the ALEKS Math test more efficient and less overwhelming.

Step 1 - Create a study plan
Step 2 - Choose your study resources
Step 3 - Review, Learn, Practice
Step 4 - Learn and practice test-taking strategies
Step 5 - Learn the ALEKS Test format and take practice tests
Step 6 - Analyze your performance

Step 1: Create a Study Plan

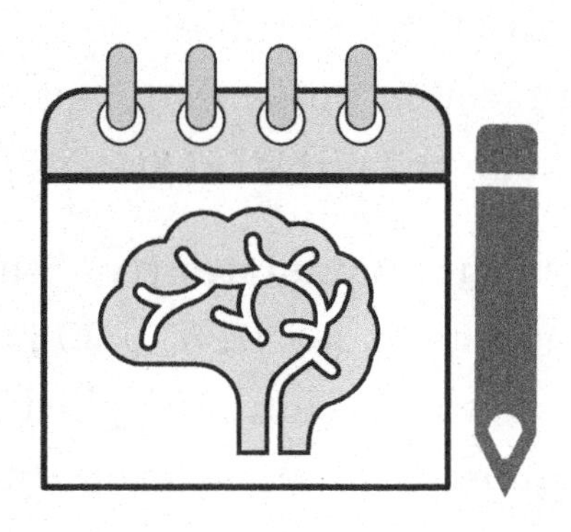

It's always easier to get things done when you have a plan. Creating a study plan for the ALEKS Math test can help you to stay on track with your studies. It's important to sit down and prepare a study plan with what works with your life, work, and any other obligations you may have. Devote enough time each day to studying. It's also a great idea to break down each section of the exam into blocks and study one concept at a time.

It's important to understand that there is no "right" way to create a study plan. Your study plan will be personalized based on your specific needs and learning style.

Follow these guidelines to create an effective study plan for your ALEKS Math test:

★ **Analyze your learning style and study habits** – Everyone has a different learning style. It is essential to embrace your individuality and the unique way you learn. Think about what works and what doesn't work for you. Do you prefer ALEKS Math prep books or a combination of textbooks and video lessons? Does it work better for you if you study every night for thirty minutes or is it more effective to study in the morning before going to work?

★ **Evaluate your schedule** – Review your current schedule and find out how much time you can consistently devote to ALEKS Math study.

★ **Develop a schedule** – Now it's time to add your study schedule to your calendar like any other obligation. Schedule time for study, practice, and review. Plan out which topic you will study on which day to ensure that you're devoting enough time to each concept. Develop a study plan that is mindful, realistic, and flexible.

★ **Stick to your schedule** – A study plan is only effective when it is followed consistently. You should try to develop a study plan that you can follow for the length of your study program.

★ **Evaluate your study plan and adjust as needed** – Sometimes you need to adjust your plan when you have new commitments. Check in with yourself regularly to make sure that you're not falling behind in your study plan. Remember, the most important thing is sticking to your plan. Your study plan is all about helping you be more productive. If you find that your study plan is not as effective as you want, don't get discouraged. It's okay to make changes as you figure out what works best for you.

Step 2: Choose Your Study Resources

There are numerous textbooks and online resources available for the ALEKS Math test, and it may not be clear where to begin. Don't worry! Effortless Math's ALEKS online center provides everything you need to fully prepare for your ALEKS Math test. In addition to the practice tests in this book, you can also use Effortless Math's online resources. (video lessons, worksheets, formulas, etc.)

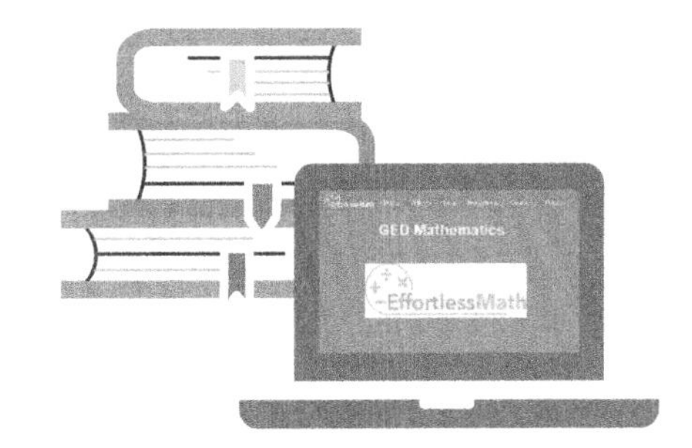

Simply visit EffortlessMath.com/ALEKS to find your online ALEKS Math resources.

STEP 3: Review, Learn, Practice

Effortless Math's ALEKS course breaks down each subject into specific skills or content areas. For instance, the percent concept is separated into different topics–percent calculation, percent increase and decrease, percent problems, etc. Use our online resources to help you go over all key math concepts and topics on the ALEKS Math test.

As you review each concept, take notes or highlight the concepts you would like to go over again in the future. If you're unfamiliar with a topic or something is difficult for you, do additional research on it. For each math topic, plenty of instructions, step-by-step guides, and examples are provided to ensure you get a good grasp of the material. You can also find video lessons on the Effortless Math website for each ALEKS Math concept.

Quickly review the topics you do understand to get a brush-up of the material. Be sure to use the worksheets and do the practice questions provided on the Effortless Math's online center to measure your understanding of the concepts.

STEP 4: Learn and Practice Test-taking Strategies

In the following sections, you will find important test-taking strategies and tips that can help you earn extra points. You'll learn how to think strategically and when to guess if you don't know the answer to a question. Using ALEKS Math test-taking strategies and tips can help you raise your score and do well on the test. Apply test taking strategies on the practice tests to help you boost your confidence.

Step 5: Learn the ALEKS Test Format and Take Practice Tests

The ALEKS *Test Review* section provides information about the structure of the ALEKS test. Read this section to learn more about the ALEKS test structure, different test sections, the number of questions in each section, and the section time limits. When you have a prior understanding of the test format and different types of ALEKS Math questions, you'll feel more confident when you take the actual exam.

Once you have read through the instructions and lessons and feel like you are ready to go – take advantage of the full-length ALEKS Math practice tests available in this book. Use the practice tests to sharpen your skills and build confidence.

The ALEKS Math practice tests offered in the book are formatted similarly to the actual ALEKS Math test. When you take each practice test, try to simulate actual testing conditions. To take the practice tests, sit in a quiet space, time yourself, and work through as many of the questions as time allows. The practice tests are followed by detailed answer explanations to help you find your weak areas, learn from your mistakes, and raise your ALEKS Math score.

Step 6: Analyze Your Performance

After taking the practice tests, look over the answer keys and explanations to learn which questions you answered correctly and which you did not. Never be discouraged if you make a few mistakes. See them as a learning opportunity. This will highlight your strengths and weaknesses.

You can use the results to determine if you need additional practice or if you are ready to take the actual ALEKS Math test.

Looking for more?

Visit EffortlessMath.com/ALEKS to find hundreds of ALEKS Math worksheets, video tutorials, practice tests, ALEKS Math formulas, and much more.

Or scan this QR code.

No Registration Required.

ALEKS Test Review

ALEKS (Assessment and Learning in Knowledge Spaces) is an artificial intelligence-based assessment tool to measure the strengths and weaknesses of a student's mathematical knowledge. ALEKS is available for a variety of subjects and courses in K-12, Higher Education, and Continuing Education. The findings of ALEKS's assessment test help to find an appropriate level for course placement. The ALEKS math placement assessment ensures students' readiness for particular math courses at colleges.

ALEKS does not use multiple-choice questions like most other standardized tests. Instead, it utilizes adaptable and easy-to-use method that mimic paper and pencil techniques. When taking the ALEKS test, a brief tutorial helps you learn how to use ALEKS answer input tools. You then begin the ALEKS Assessment. In about 30 to 45 minutes, the test measures your current content knowledge by asking 20 to 30 questions. ALEKS is a Computer Adaptive (CA) assessment. It means that each question will be chosen on the basis of answers to all the previous questions. Therefore, each set of assessment questions is unique. The ALEKS Math assessment does not allow you to use a personal calculator. But for some questions ALEKS onscreen calculator button is active and the test taker can use it.

Key Features of the ALEKS Mathematics Assessment

Some key features of the ALEKS Math assessment are:

- ❖ Mathematics questions on ALEKS are adaptive to identify the student's knowledge from a comprehensive standard curriculum, ranging from basic arithmetic up to precalculus, including trigonometry but not calculus.
- ❖ Unlike other standardized tests, the ALEKS assessment does not provide a "grade" or "raw score." Instead, ALEKS identifies which concepts the student has mastered and what topics the student needs to learn.
- ❖ ALEKS does not use multiple-choice questions. Instead, students need to produce authentic mathematical input.
- ❖ There is no time limit for taking the ALEKS Math assessment. But it usually takes 30 to 45 minutes to complete the assessment.

The ALEKS Math score is between 1 and 100 and is interpreted as a percentage correct. A higher ALEKS score indicates that the test-taker has mastered more math concepts. ALEKS Math assessment tool evaluates mastery of a comprehensive set of mathematics skills ranging from basic arithmetic up to precalculus, including trigonometry but not calculus. It will place students in classes up to Calculus.

ALEKS Math Test-Taking Strategies

Here are some test-taking strategies that you can use to maximize your performance and results on the ALEKS Math test.

#1: Use This Approach To Answer Every ALEKS Math Question

- Review the question to identify keywords and important information.
- Translate the keywords into math operations so you can solve the problem.
- Review the answer choices. What are the differences between answer choices?
- Draw or label a diagram if needed.
- Try to find patterns.
- Find the right method to answer the question. Use straightforward math, plug in numbers, or test the answer choices (backsolving).
- Double-check your work.

#2: Answer Every ALEKS Math Question

Don't leave any fields empty! ALEKS is a Computer Adaptive (CA) assessment. Therefore, you cannot leave a question unanswered and you cannot go back to previous questions.

Even if you're unable to work out a problem, strive to answer it. Take a guess if you have to. You will not lose points by getting an answer wrong, though you may gain a point by getting it correct!

#3 : Ballpark

A ballpark answer is a rough approximation. When we become overwhelmed by calculations and figures, we end up making silly mistakes. A decimal that is moved by one unit can change an answer from right to wrong, regardless of the number of steps that you went through to get it. That's where ballparking can play a big part.

If you think you know what the correct answer may be (even if it's just a ballpark answer), you'll usually have the ability to estimate the range of possible answers and avoid simple mistakes.

#4 : Plugging In Numbers

"Plugging in numbers" is a strategy that can be applied to a wide range of different math problems on the ALEKS Math test. This approach is typically used to simplify a challenging question so that it is more understandable. By using the strategy carefully, you can find the answer without too much trouble.

The concept is fairly straightforward–replace unknown variables in a problem with certain values. When selecting a number, consider the following:

- Choose a number that's basic (just not too basic). Generally, you should avoid choosing 1 (or even 0). A decent choice is 2.
- Try not to choose a number that is displayed in the problem.
- Make sure you keep your numbers different if you need to choose at least two of them.
- If your question contains fractions, then a potential right answer may involve either an LCD (least common denominator) or an LCD multiple.
- 100 is the number you should choose when you are dealing with problems involving percentages.

ALEKS Mathematics – Test Day Tips

After practicing and reviewing all the math concepts you've been taught, and taking some ALEKS mathematics practice tests, you'll be prepared for test day. Consider the following tips to be extra-ready come test time.

Before Your Test

What to do the night before:

- Relax! One day before your test, study lightly or skip studying altogether. You shouldn't attempt to learn something new, either. There are plenty of reasons why studying the evening before a big test can work against you. Put it this way–a marathoner wouldn't go out for a sprint before the day of a big race. Mental marathoners–such as yourself–should not study for any more than one hour 24 hours before a ALEKS test. That's because your brain requires some rest to be at its best. The night before your exam, spend some time with family or friends, or read a book.
- Avoid bright screens - You'll have to get some good shuteye the night before your test. Bright screens (such as the ones coming from your laptop, TV, or mobile device) should be avoided altogether. Staring at such a screen will keep your brain up, making it hard to drift asleep at a reasonable hour.
- Make sure your dinner is healthy - The meal that you have for dinner should be nutritious. Be sure to drink plenty of water as well. Load up on your complex carbohydrates, much like a marathon runner would do. Pasta, rice, and potatoes are ideal options here, as are vegetables and protein sources.
- Get your bag ready for test day – Prefer to take ALEKS in the Testing Office? The night prior to your test, pack your bag with your stationery, admissions pass, ID, and any other gear that you need. Keep the bag right by your front door. If you prefer to take the test at home, find a quite place without any distractions.
- Make plans to reach the testing site – If you are taking the test at the testing office, ensure that you understand precisely how you will arrive at the site of the test. If parking is something you'll have to find first, plan for it. If you're dependent on public transit, then review the schedule. You should also make sure that the train/bus/subway/streetcar you use will be running. Find out about road closures as well. If a parent or friend is accompanying you, ensure that they understand what steps they have to take as well.

The Day of the Test

- **Get up reasonably early, but not too early.**
- **Have breakfast** - Breakfast improves your concentration, memory, and mood. As such, make sure the breakfast that you eat in the morning is healthy. The last thing you want to be is distracted by a grumbling tummy. If it's not your own stomach making those noises, another test taker close to you might be instead. Prevent discomfort or embarrassment by consuming a healthy breakfast. Bring a snack with you if you think you'll need it.
- **Follow your daily routine** - Do you watch TV in the morning while getting ready for the day? Don't break your usual habits on the day of the test. Likewise, if coffee isn't something you drink in the morning, then don't take up the habit hours before your test. Routine consistency lets you concentrate on the main objective–doing the best you can on your test.
- **Wear layers** - Dress yourself up in comfortable layers if you are taking the test at the testing site. You should be ready for any kind of internal temperature. If it gets too warm during the test, take a layer off.
- **Make your voice heard** - If something is off, speak to a proctor. If medical attention is needed or if you'll require anything, consult the proctor prior to the start of the test. Any doubts you have should be clarified. You should be entering the test site with a state of mind that is completely clear.
- **Have faith in yourself** - When you feel confident, you will be able to perform at your best. When you are waiting for the test to begin, envision yourself receiving an outstanding result. Try to see yourself as someone who knows all the answers, no matter what the questions are. A lot of athletes tend to use this technique–particularly before a big competition. Your expectations will be reflected by your performance.

During your test

- **Be calm and breathe deeply** - You need to relax before the test, and some deep breathing will go a long way to help you do that. Be confident and calm. You got this. Everybody feels a little stressed out just before an evaluation of any kind is set to begin. Learn some effective breathing exercises. Spend a minute meditating before the test starts. Filter out any negative thoughts you have. Exhibit confidence when having such thoughts.
- **Concentrate on the test** - Refrain from comparing yourself to anyone else. You shouldn't be distracted by the people near you or random noise. Concentrate exclusively on the test. If you find yourself irritated by surrounding noises, earplugs can be used to block sounds off close to you. Don't forget–the test is going to last an hour or more. Some of that time will be dedicated to brief sections. Concentrate on the specific section you are working on during a particular moment. Do not let your mind wander off to upcoming or previous questions.
- **Try to answer each question individually** - Focus only on the question you are working on. Use one of the test-taking strategies to solve the problem. If you aren't able to come up with an answer, don't get frustrated. Simply guess, then move onto the next question.
- **Don't forget to breathe!** Whenever you notice your mind wandering, your stress levels boosting, or frustration brewing, take a thirty-second break. Shut your eyes, drop your pencil, breathe deeply, and let your shoulders relax. You will end up being more productive when you allow yourself to relax for a moment.

After your test

- **Take it easy** - You will need to set some time aside to relax and decompress once the test has concluded. There is no need to stress yourself out about what you could've said, or what you may have done wrong. At this point, there's nothing you can do about it. Your energy and time would be better spent on something that will bring you happiness for the remainder of your day.
- **Redoing the test** - Did you succeed on the test? Congratulations! Your hard work paid off! Succeeding on this test means that you are now ready to take college level courses. If you didn't receive the result you expected, though, don't worry! The test can be retaken. In such cases, you will need to follow the retake policy. You also need to re-register to take the exam again.

Contents

Time to Test

Time to refine your skills with a practice test.

In this book, there are 10 complete ALEKS Mathematics Tests. Take these tests to simulate the test day experience. After you've finished, score your test using the answers and explanations section.

Before You Start

- You'll need a pencil, a scientific calculator, and scratch papers to take the test.
- For these practice tests, don't time yourself. Spend time as much as you need.
- After you've finished the test, review the answer key to see where you went wrong.

Good Luck

ALEKS Mathematics

Practice Test 1

2023 - 2024

Total number of questions: 35

Total time: No time limit

Calculators are permitted for ALEKS Math Test.

1) What value of x makes the following equation true? $\frac{48}{5} = \frac{6}{x-1}$

2) Write values for x and y to satisfy the following system of equations.

$$\begin{cases} x + 4y = 10 \\ 5x + 10y = 20 \end{cases}$$

3) At a store, 8 notebooks cost $16. Each notebook has the same cost. Write an equation that represents the cost of each notebook at this store.

4) If $\frac{x-3}{5} = N$ and $N = 6$, what is the value of x?

5) John has completed 18 out of 25 problems in his physics homework. What percentage of the problems has John not yet completed?

6) If $x + y = 7$ and $x^2 + y^2 = 9$, find xy.

7) Simplify $3i(2i - 3) + (1 - 5i)(2i)$.

8) What is the following series in sigma notation?

$$-1 + \frac{r}{4} - \frac{r^2}{7} + \frac{r^3}{10} - \frac{r^4}{13} + \cdots + \frac{(-1)^{n+1}}{3n+1} r^n + \cdots$$

9) What is the sum of the solutions to $(3 - x)(x + 0.3) = 0$?

10) Quadratic functions g and h are shown below.

$$g(x) = (x - c)^2$$
$$h(x) = x^2 + 2x - 1$$

For what value of c will the graph of h be 2 units below the graph of g?

x	1	2	3
$g(x)$	-1	-3	-5

11) The table above shows some values of linear function $g(x)$. Find $g(x)$.

12) f is an exponential function defined by $f(x) = a2^{bx-1} - 1$, where a and b are positive constants. If $f(1) = 1$ and $f(2) = 15$, what is the value of $a + b$?

13) What is the y −intercept of the line with the equation $x - 3y = 12$?

14) Simplify $\sqrt[3]{92^4}\left(\frac{128}{3^{-1}\sqrt[4]{16}}\right)^{\frac{1}{3}}$.

15) The price of chocolate was raised from \$5.40 to \$5.67. What was the percent increase in the price?

16) A candy company produces and sells cases of chocolate bars. The company incurs fixed costs of $21,000, and it costs an additional $4 to make each case of chocolate bars. The selling price for each case is $10. The graph of the system of linear equations representing the company's costs and revenue for manufacturing and selling x cases of chocolate bars is shown below.

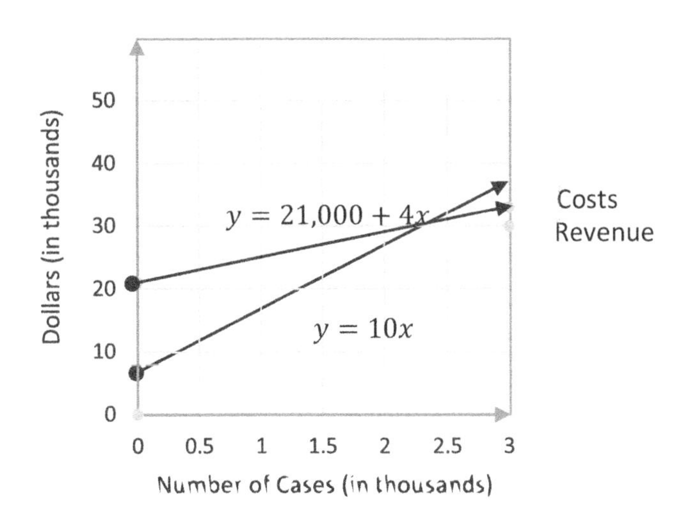

What is the number of cases of chocolate bars the company needs to sell to achieve equal costs and revenue?

17) The approximate distance in kilometers between two cities can be found using the function $k = 60t + 180$, where t is the number of hours a car has been traveling. What number of hours and minutes is closest to the amount of time that the car has been traveling if the distance between the two cities is 500 kilometers?

18) What is the value of $\frac{|1-2(4+1)|}{|-(3-15)|} - 7$?

19) The locust population in a small plain grows exponentially. The current population is 243 locusts, and the relative growth rate is $\frac{3}{2}$ per week. Find the average rate of change of the locust population between the 4th and 12th weeks of the year.

20) What is the sum of the first 5 terms of the series?

$$2, 8, 64, \cdots$$

21) Tickets to a movie cost $12.50 for adults and $7.50 for students. A group of 12 friends purchased tickets for $125. How many student tickets did they buy?

22) For what value of x is the proportion true? $x: 40 = 20: 32$

23) A company has a budget of at most \$10,000 to purchase new laptops and printers. Each laptop costs \$800 and each printer costs \$250. Which inequality represents all possible combinations of x, the number of laptops, and y, the number of printers, the company can buy while staying within its budget?

24) Jack earns \$616 for his first 44 hours of work in a week and is then paid 1.5 times his regular hourly rate for any additional hours. This week, Jack needs \$826 to pay his rent, bills, and other expenses. How many hours must he work to make enough money this week?

25) In an arithmetic sequence, the 2nd sequence is 7 and the 8th sequence is 31, what is the sum of the first 20 sequences of this series?

26) In a school of 120 students, 40 students are in the theater group, 60 students participate in the football team, and 20 students participate in both activities. How many students in the school are neither in the theater group nor on the football team?

27) Let the sets A and B be as follows:

$A = \{x \in \mathbb{N}: 3 \leq x < 15 \text{ and } x \text{ is an even number}\}$,

$B = \{x \in \mathbb{N}: 7 < x \leq 30 \text{ and } x \text{ is multiple of } 7\}$.

How many elements are in set $A \cup B$?

28) Anna opened an account with a deposit of $3,000. This account earns 5% simple interest annually. How many years will it take her to earn $600 on her $3,000 deposit?

29) What is the 15th term of the arithmetic sequence $\frac{1}{2}x, \frac{1}{2}x + 3, \frac{1}{2}x + 6, \ldots$?

30) There are 97 students in a classroom. There are 15 more girls than boys. What is the total number of girls in the classroom?

31) Let the remainder of the polynomial $P(x) = x^N + A$ divided by $x + 2$ and $x + 1$ be 2 and -1 respectively. Find factor $P(x)$.

32) A function is shown:

$$g(x) = 3x - 10$$

What is the value of $g(4)$?

33) The first three terms of a geometric sequence are 6, 9, and 13.5. What is the nth term of the sequence?

34) Find the 5th entry in row 9th of Pascal's triangle.

35) What is the biggest value of x in the following equation?

$$x(x^2 - 1)^2 - 8x^3 = -8x$$

This is the end of Practice Test 1.

ALEKS Mathematics

Practice Test 2

2023 - 2024

Total number of questions: 35

Total time: No time limit

Calculators are permitted for ALEKS Math Test.

1) Expand $(3x - 5)^2$.

2) $(2i - 3) - (2 - i)(i - 2) =$

3) $\frac{3x-2}{x-3} - \frac{x-4}{3x-2} =$

4) If $2x - 5y = 10$, what is x in terms of y?

5) If $x = \frac{1}{3}$ and $y = \frac{9}{21}$, then what is equal to $\frac{1}{x} \div \frac{y}{3}$?

6) Given that $(n - x)^2 \left(\frac{1}{2}x + 1\right)^n = 9 + \frac{15}{2}x + \cdots$, find the value of n.

7) Solve for: $-4 \leq 4x - 8 < 16$.

8) If $2x - 4 = 14$, what is the value of $3x - 3$?

9) For $i = \sqrt{-1}$, what is the value of $\frac{3+2i}{5+i}$?

10) How many terms are there in a geometric series if the first term is 5, the common ratio is 2, and the sum of the series is 1275?

11) Find the value of x, when $C(x, 3) = P(x, 3)$.

12) If $f(x^2) = 3x + 4$, for all positive values of x, what is the value of $f(121)$?

13) The line k is parallel to the line $y = \frac{3}{4}x + 3$ and intersects the y −axis at point −7. If point $m(12, b)$ is on the line k, what is the value of b?

14) If a, b, and c are positive integers and $3a = 4b = 5c$, then the value of $a + 2b + 15c$ is how many times the value of a?

15) The formula below is used in medicine to estimate the body surface area A, in square meters, of infants and children whose weight w ranges between 3 and 30 kilograms and whose height h is measured in centimeters. Based on the current formula, what is w in terms of A?

Current's formula: $A = \frac{4+w}{30}$

16) If $h^{\frac{k}{6}} = 125$ for positive integers h and k, what is the smallest possible value subtracted from the largest possible value for k?

17) To buy a new computer, Emma borrowed \$2,500 at 8% interest for 6 years. How much interest did she pay?

18) What is the inverse of $y = x^2 - 2x$?

19) If $log_3(x^2 + 5) = 2$, then $x =$?

20) If $8 + 2x$ is 16 more than 20, what is the value of $6x$?

21) If $log_x 2 = \frac{1}{3}$, what is the value of x?

22) Find the equation of the horizontal asymptote of the function $f(x) = \frac{x+2}{x^2+1}$.

23) The equation of the function $f(x) = (x - 1)(x^2 - 4)$ is given. Write an equation for a translation of $f(x)$ that has no x −intercept.

24) Solve the following inequality: $\left|\frac{x}{2} - 2x + 10\right| < 5$.

25) What is the value of x in this equation?

$$4\sqrt{2x + 9} = 28$$

26) A container holds 3.5 gallons of water when it is $\frac{7}{24}$ full. How many gallons of water does the container hold when it's full?

27) What is the equation of the following graph?

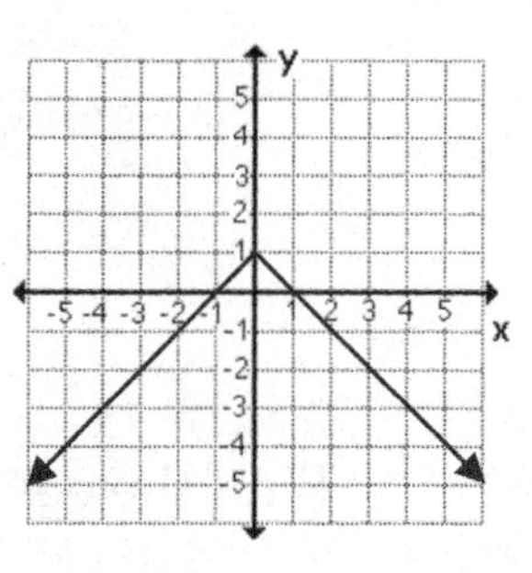

28) What is the coefficient x^7 of $\sum_{k=1}^{n} \frac{k-1}{k!} x^{2k-1}$?

29) What is the sum of $\sqrt{x-7}$ and $\sqrt{x}-7$ when $\sqrt{x}=4$?

30) Determine the intervals where the function is positive and negative.

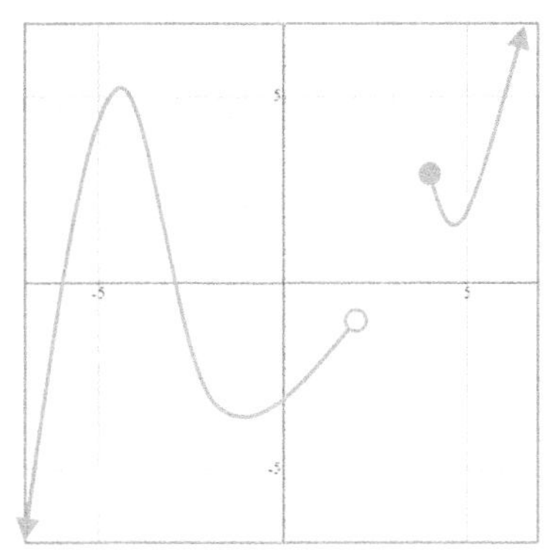

31) What are the domain and range of this graph?

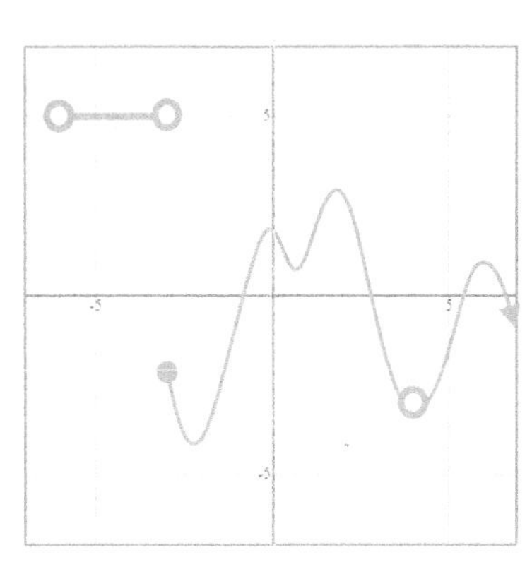

32) Solve for x: $\frac{3x}{5}=27$.

33) What is the solution of the equation $81 = 3^{2x}$?

34) A senior employee who works 20 hours earns $28.00 more than a junior employee who works 15 hours. The senior employee earns $17 per hour. What is the hourly pay in dollars and cents for the junior employee?

35) Sara opened a bank account that earns 2 percent compounded annually. Her initial deposit was $150, and she uses the expression $\$150(x)^n$ to find the value of the account after n years. What is the value of x in the expression?

This is the end of Practice Test 2.

ALEKS Mathematics

Practice Test 3

2023 - 2024

Total number of questions: 35

Total time: No time limit

Calculators are permitted for ALEKS Math Test.

1) $(x^6)^{\frac{7}{8}}$ is equal to …

2) The functions f and g are defined below. What is the value of $f(g(-2))$?

$$f(x) = -x + 4$$
$$g(x) = |2x|$$

3) Mr. Brendan weighed 80 kg in 2003 and weighs 170 kg in 2021. What was the rate of change in his weight?

4) If Anna multiplies her age by 5 and then adds 3, she will get a number equal to her mother's age. If x is her mother's age, what is Anna's age in terms of x?

5) What is the value of x in the geometric sequence $\left\{x, \frac{1}{2}, -\frac{1}{8}, \cdots\right\}$?

6) $i(4-5i)+i-(2i+3)=?$

7) What is the value of x in the following equation? $\frac{3}{4}(x-2)=3(\frac{1}{6}x-\frac{3}{2})$

8) In field research, a student surveyed the members of a public library and recorded the interest of the library members in the three subjects of philosophy, history, and novels. He made the Venn diagram below to show the members' interest in each of the three areas: Set P for philosophy, set H for history, and N for novels.

How many members belong to the set $(H \cap N) \cup P$?

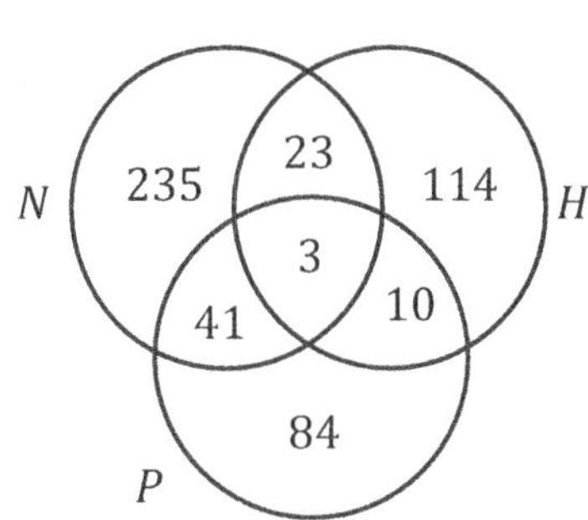

9) If $4n-3 \geq 1$, what is the least possible value of $4n+3$?

10) For what real value of x is the equation below true?

$$x^3-6x^2+3x-18=0$$

11) What is the slope of the line shown?

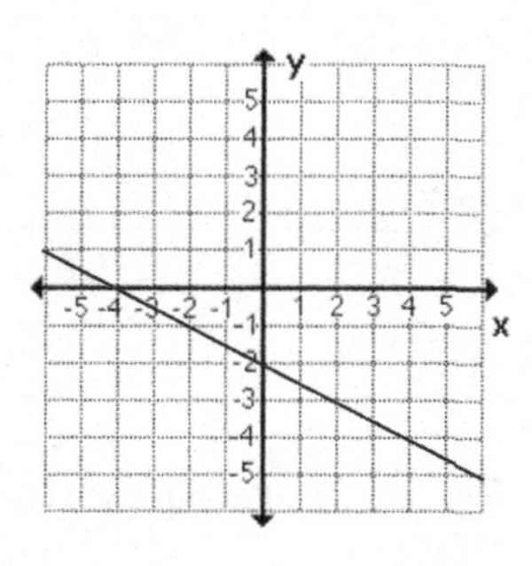

12) Last week 18,000 fans attended a football match. This week three times as many bought tickets, but one-sixth of them canceled their tickets. How many are attending this week?

13) What is the sum of the first five sentences at the beginning of the series $\sum_{i=1}^{n}(i^2 + 3i - 4)$?

14) Solve the following system of equations.

$$3y + 4x = 3x^2 + 2$$
$$x + 2 = 3y - 2(x - 1)^2$$

15) A group of hikers is planning to climb a mountain that is 5,000 feet high. Which inequality can be used to find all possible values of t, the time it will take the hikers to climb the mountain in hours, if they climb at an average speed of at least r feet per hour?

$$f(x) = \frac{1}{(x-3)^2 + 4(x-3) + 4}$$

16) For what value of x is the function $f(x)$ above undefined?

17) How long will it take to receive $360 from an investment of $240 at the rate of 10% simple interest?

18) What is the inverse of the function $f(x) = 5 \cdot \left(\frac{1}{2}\right)^x$?

19) $f(a) = |11 + a^2|$, where a is a positive integer. If $f(a) = 20$, what is the value of a that satisfies the equation above?

20) x is $y\%$ of what number?

21) The sum of four numbers is 600. One of the numbers, x is 50% more than the sum of the other three numbers. What is the value of x?

22) The profit in dollars from a carwash is given by the function $P(x) = \frac{40a-500}{a} + b$, where a is the number of cars washed and b is a constant. If 50 cars were washed today for a total profit of $600, what is the value of b?

23) Consider the expansion of $x\left(2x - \frac{c}{x}\right)^5$. The constant term is 5. Find c.

24) Perform the operations and simplify: $5\sqrt{2} + \sqrt{32} - 2^{\frac{1}{2}}$.

25) In the following equation, when z is divided by 3, what is the effect on x?

$$x = \frac{8y + \frac{r}{r+1}}{\frac{6}{z}}$$

26) If $a = b + 2c$ and $b = c$, then:

27) Solve this equation for $x = 0$: $f(x) = 2^{3x-1} + 1$.

28) Find the value of y in the following system of equations.

$$3x - 4y = -20$$
$$-x + 2y = 10$$

29) If $f(x) = \frac{1}{2}x + 2$, then $f(2x - 4) =$?

30) If $\frac{1}{b-1} = \frac{1}{c+2}$, then $c =$?

31) If $0.00104 = \frac{104}{x}$, what is the value of x?

32) $\frac{1\frac{4}{3}+\frac{1}{4}}{2\frac{1}{2}-\frac{17}{8}}$ is approximately equal to:

33) What is the vertical asymptote of the graph $y = \frac{4x-2}{3x+5}$?

34) Jason drew Pascal's triangle with m rows. What is the total number of entries?

35) What is the domain of the function shown below?

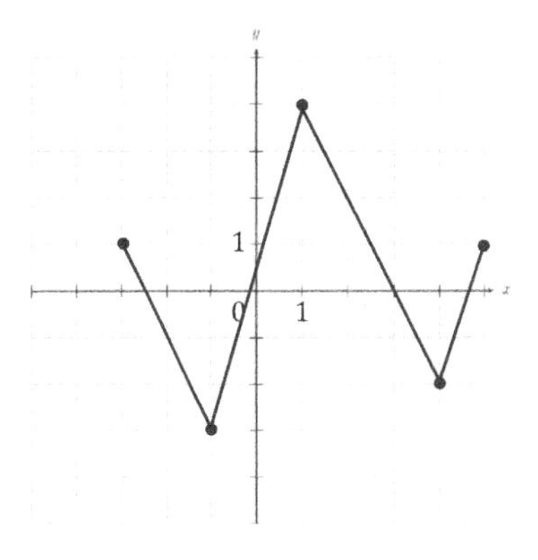

This is the end of Practice Test 3.

ALEKS Mathematics

Practice Test 4

2023 - 2024

Total number of questions: 35

Total time: No time limit

Calculators are permitted for ALEKS Math Test.

1) If $f(x) = 3x + 4(x + 1) + 2$, then $f(4x) =?$

2) Simplify $(-4 + 9i)(3 + 5i)$.

3) Simplify this expression: $\frac{(-2x^2y^2)^3(3x^3y)}{12x^3y^8}$.

4) Solve and write the solution set the in set-builder notation of the following inequality.

$$2x - 4(x + 2) \geq x + 6$$

5) What is the negative solution to $2x^3 - x^2 - 6x = 0$?

6) The table shows the linear relationship between the profit earned in million dollars at a home appliance factory and the number of sales of a product.

Number of Sales of a Product per thousand	1	3	7	14	15
The Profit earned in Million dollars	13	21	37	65	69

What is the rate of change in profit obtained in a million dollars with respect to the number of product sales in the factory?

7) What is the zero of $r(x) = \frac{4}{7}x + 12$?

8) The sum of $-4x^2 + 3x - 24$ and $7x^2 - 8x + 18$ can be written in the form $ax^2 + bx + c$, where a, b, and c are constants. What is the value of $a + b - c$?

9) What are the zeros of the function: $f(x)=x^2 - 7x + 12$?

10) Find the factors of the binomial $x^3 - 8$.

11) What is the value of the y –intercept of the graph of $k(x) = 42\left(\frac{4}{5}\right)^x$?

12) If $3^m . 9^n = 3^{12}$, what is the value of $m + 2n$?

13) A circle is inscribed in a square and the radius of the circle is 4. What is the area of the shaded region?

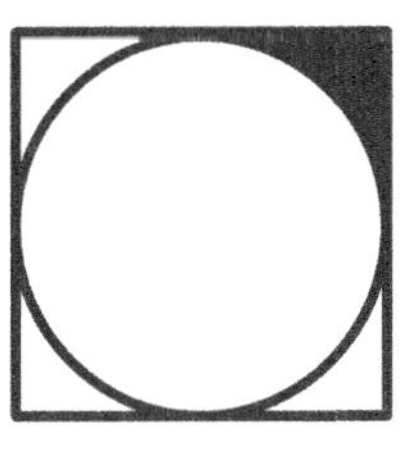

14) Given $q(x) = 3(x - 5)^2 - 7$, what is the value of $q(2)$?

15) $(7x + 2y)(5x + 2y) =?$

16) Multiply and write the product in scientific notation:

$$(2.9 \times 10^6) \times (2.6 \times 10^{-5})$$

17) Find $-2 + \frac{3b-4c}{9b} - \frac{2b+2c}{6b}$.

18) What is the number of solutions to the equation $x^2 - 3x + 1 = x - 3$?

19) Multiply and write the product in scientific notation:

$$(1.7 \times 10^9) \times (2.3 \times 10^{-7})$$

$$h = -25t^2 + st + k$$

20) The equation above gives the height h, in feet, of a ball t seconds after it is thrown straight up with an initial speed of s feet per second from a height of k feet. Find s in terms of h, t, and k.

21) Solve: $2x(5 + 3y + 2x + 4z)$.

22) What is the value of x in the following equation?

$$log_4(x+2) - log_4(x-2) = 1$$

23) A customer purchased movie tickets online. The total cost, c, in dollars, of t tickets can be found using the function below.

$$c = 24.50t + 5.25$$

If the customer spent a total of $103.25 on tickets, how many tickets did he purchase?

24) Mr. Anderson has a beaker containing n milliliters of a solution to distribute to the students in his chemistry class. If he gives each student 3 milliliters of the solution, he will have 5 milliliters left over. In order to give each student 4 milliliters of the solution, he will need an additional 21 milliliters. How many students are in the class?

25) Find the x –intercept and y –intercept: $3x - 6y = 24$.

26) Approximately, what is the perimeter of the figure below? ($\pi = 3$)

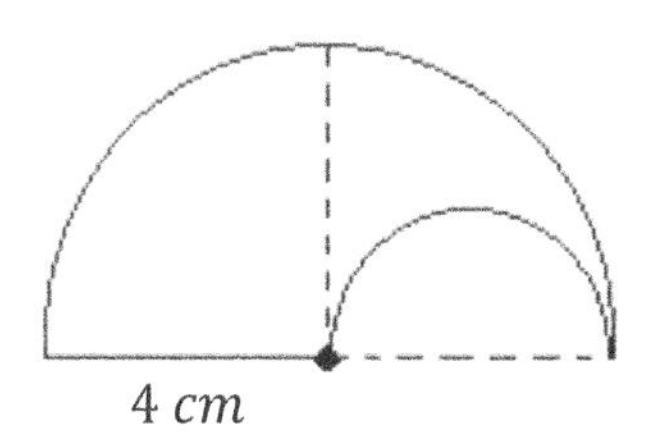

27) What is the inverse of the function $f(x) = x^2 + 1$?

28) Perform the operations and simplify $\sqrt{8} - \sqrt{50} + \sqrt{72}$.

$$y = x^2 - 9x + 18$$

29) The equation above represents a parabola in the xy −plane. What is one x −intercept of the parabola?

30) What is the solution to the system of equations below?

$$4x - 7y = -2$$
$$12x - 21y = -42$$

31) What is the domain of $f(x) = -4x^2 + 25$?

32) The exponential growth $f(x) = 5(3)^x$ is shown in the following table. Find the average rate of change over the given interval $1 \leq x \leq 3$.

x	0	1	2	3	4
$f(x)$	5	15	45	135	450

33) Find $\sum_{i=1}^{n}(i+2)$.

34) According to the graph, what is the minimum number of degrees of the function $f(x)$?

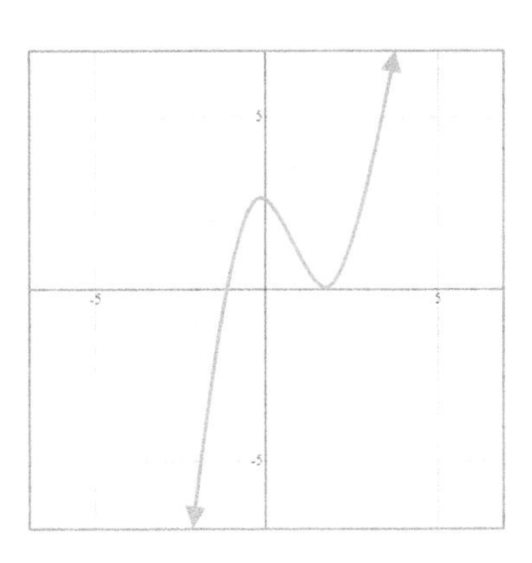

35) The table is given below for the continuous function $f(x)$. What is the minimum degree for the function?

x	$f(x)$
-3	8
-1	2
0	-1
2	14
5	-7
10	-3
18	1

This is the end of Practice Test 4.

ALEKS Mathematics

Practice Test 5

2023 - 2024

Total number of questions: 35

Total time: No time limit

Calculators are permitted for ALEKS Math Test.

1) At a rate of $3d + 9$ Kilometers per hour, how many kilometers can a train travel in 8 hours?

2) The regular price for a concert ticket is $100. A different ticket vendor offers a 10% discount on the regular price. What would be the savings in dollars and cents if you purchase 3 tickets from the discounted vendor instead of the regular vendor?

3) Solve: $\frac{1.4\times10^{-7}}{2\times10^{-10}}$.

4) If y varies directly with x, the relationship can be represented by the equation $y = kx$, where k is the proportionality constant. Given that $y = 6$ when $x = 24$, what is the equation of the direct variation that represents this relationship?

5) If $x = 3\left(log_2 \frac{1}{32}\right)$, what is the value of x?

6) The table represents different values of function $g(x)$. What is the value of the expression $3g(-2) - 2g(3)$?

x	$g(x)$
-2	3
-1	2
0	1
1	0
2	-1
3	-2

7) If a quadratic function with equation $y = ax^2 + 5x + 10$, where a is constant, passes through the point $(2, 12)$, what is the value of a^2?

8) The graph of quadratic function f is shown on the grid.

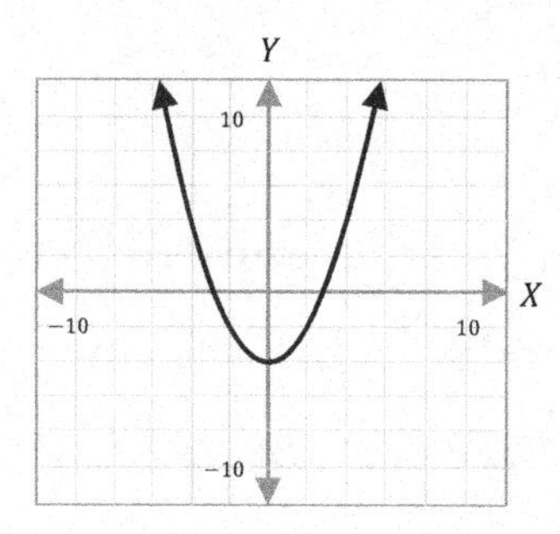

If $g(x) = x^2$ and $f(x) = g(x) + k$, what is the value of k?

$$3x + x + x - 2 = x + x + x + 8$$

9) In the equation above, what is the value of x?

10) The area of the following equilateral triangle with sides of length d is …

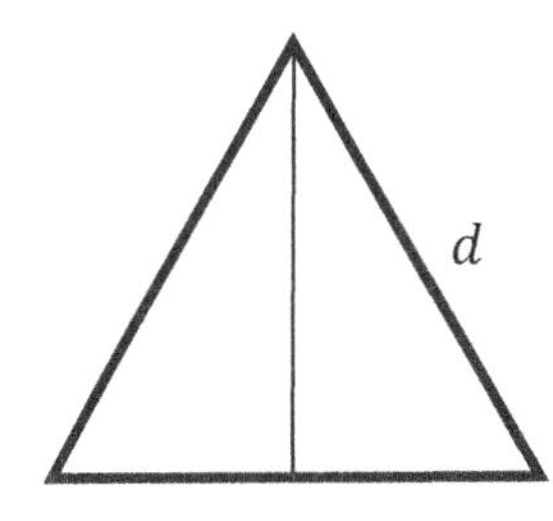

11) What is the solution to the following inequality? $|x - 9| \leq 5$

12) Express in scientific notation: 0.00095.

13) If $x^2 + \frac{1}{x^2} = 7$, find the value of $x^4 + \frac{1}{x^4}$.

14) Find a factor of $6x^2 - 4x - 10$.

15) The expression $(xy^{-2})^3 \left(\frac{y}{x}\right)^9$ is equivalent to $x^n y^m$. What is the value of $n - m$?

16) Solve: $4x - 3y < 2y + 35$.

17) For what value of x is the function $f(x)$ below undefined?

$$f(x) = \frac{1}{(x-3)^2 - 4}$$

18) What is the sum of all values of n that satisfies $2n^2 + 16n + 24 = 0$?

19) A construction worker can complete building a brick wall in 5 hours and a wooden fence in 3 hours. The function below can be used to find the number of brick walls the worker builds when she completes f wooden fences in a 40-hour workweek.

$$b = \frac{(50 - 3f)}{0.5}$$

If the worker built 10 brick walls in one week, how many wooden fences did she complete that week?

20) Subtract $5x^2 - 3$ from triple the quantity $-x^2 - 2x + 2$.

21) Simplify $(3x - 5)^2$.

22) Solve: $f(x) = 2(2 - 3x)^2 - 7$.

23) In the following figure, point O is the center of the circle and the equilateral triangle has a perimeter of 45. What is the circumference of the circle? ($\pi = 3$)

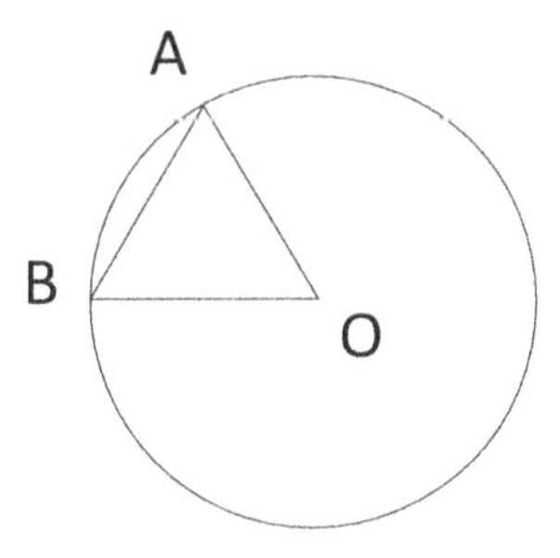

24) $(5x - 3y)^2$.

25) Solve this equation for x: $e^{3x} = 18$.

26) Simplify $\frac{4-3i}{-4i}$.

27) What is the solution to $2(n-1) = 3(n+2) - 10$?

28) If a and b are solutions of the following equation, what is the ratio $\frac{a}{b}$? $(a > b)$

$$2x^2 - 11x + 8 = -3x + 18$$

29) Point A lies on the line with equation $y - 3 = 2(x + 5)$. If the x −coordinate of A is 8, what is the y −coordinate of A?

30) The graph of $y = -3x^2 + 12x + 6$ is shown below. If the graph crosses the y −axis at the point $(0, r)$, what is the value of r?

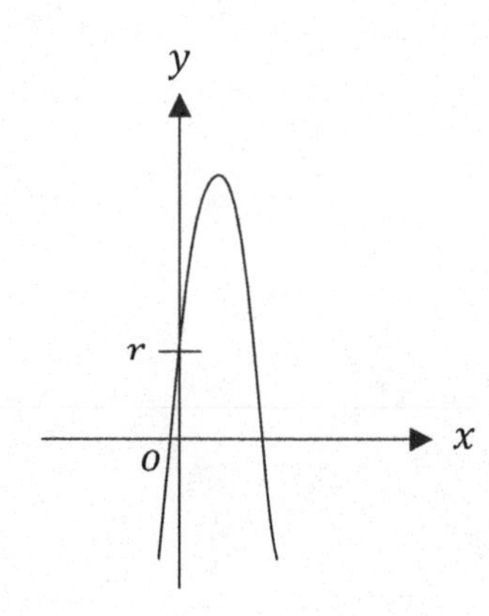

31) What is the range of the function graphed on the grid?

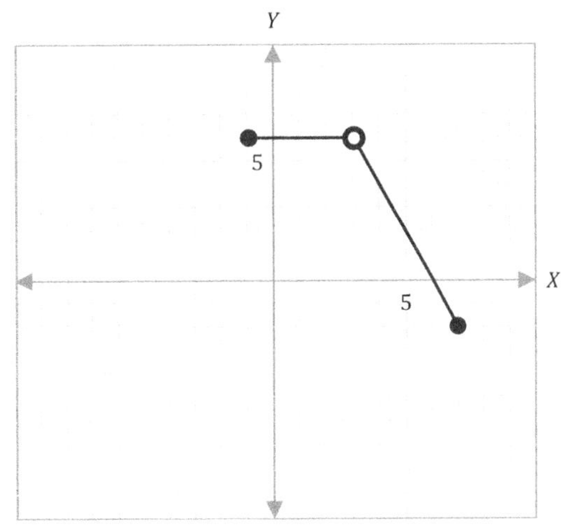

32) What expression is equivalent to $-49x^2 + 9$?

33) If $f(x) = 3x + 4(x + 1) + 2$, then $f(3x) =$?

34) If $f(x) = \frac{10x-3}{6}$ and $f^{-1}(x)$ is the inverse of $f(x)$, what is the value of $f^{-1}(2)$?

35) The perimeter of a rectangular garden is 38 meters. The length of the garden can be represented by $(x + 6)$ meters, and its width can be represented by $(2x - 2)$ meters. What are the dimensions of this garden in meters?

This is the end of Practice Test 5.

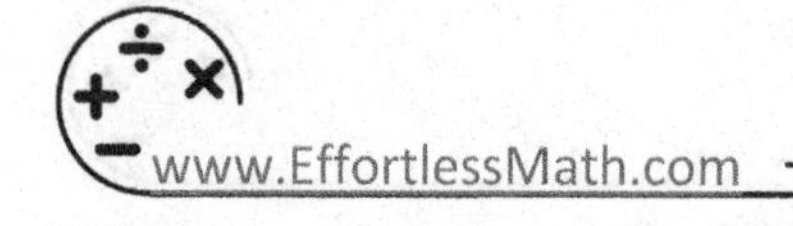

ALEKS Mathematics

Practice Test 6

2023 - 2024

Total number of questions: 35

Total time: No time limit

Calculators are permitted for ALEKS Math Test.

1) A website for home appliance repair services charges businesses a one-time setup fee of $100 plus x dollars for each quarter (every 3 months). If a business owner paid $260 for the first 3 months, including the setup fee, what is the value of x?

2) Use a calculator to approximate $\sqrt{508}$.

   Round your answer to the nearest hundredth.

3) If $e^{ln4} = x$, what is the value of x?

4) Solve for a.

   $-3a + 8(a + 8) = 49$

   Simplify your answer as much as possible.

5) What is the solution to $1 = \frac{6x}{3+x} - x$?

6) Sketch the graph of inequality: $-8x < 16 - 4y$

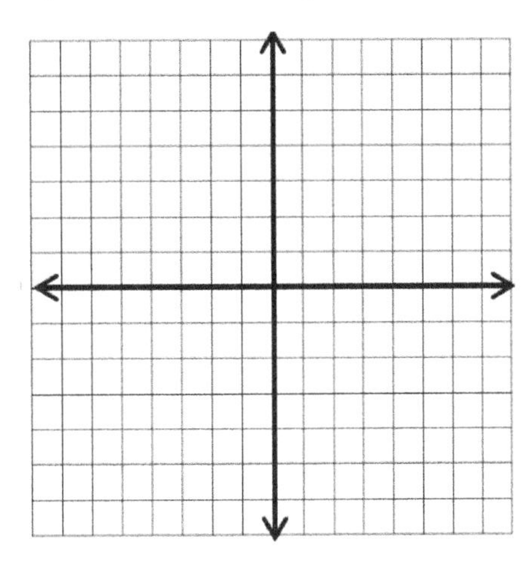

7) Simplify $(-5 + 10i)(-3 - 13i)$.

8) What is the perimeter of the following trapezoid?

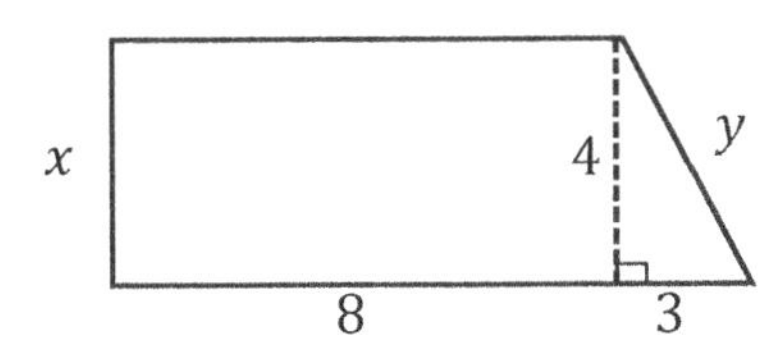

9) What is the value of y in the following system of equations?

$$3x - 4y = -20$$
$$-x + 2y = 20$$

10) How long does a 396–miles trip take moving at 55 miles per hour (mph)?

11) When 50% of 60 is added to 12% of 600, the resulting number is:

12) What is the interval solution to the inequality $\frac{6x+12}{x-8} > 0$?

13) A bag contains 18 balls: two green, five black, eight blue, a brown, a red, and one white. If 17 balls are removed from the bag at random, what is the probability that a brown ball has been removed?

14) The complete graph of the function f is shown in the xy −plane above. For what value of x is the value of $f(x)$ at its minimum?

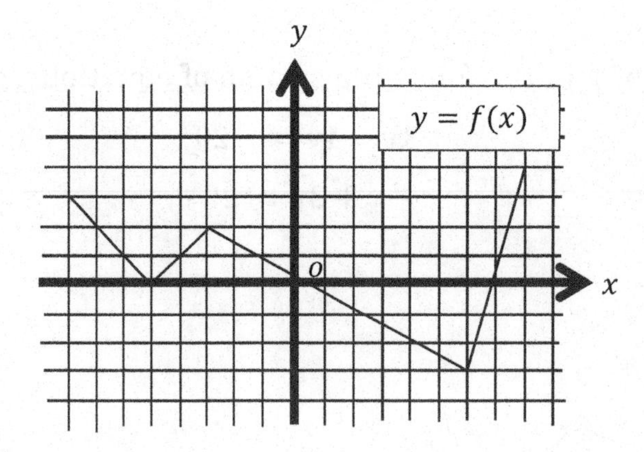

15) What is the value of $sin\ 45°$?

16) Simplify. $2x^2 + 4y^5 - x^2 + 2z^3 - 2y^2 + 2x^3 - 2y^5 + 6z^3$

17) A football team had \$20,000 to spend on supplies. The team spent \$14,000 on new balls. New sports shoes cost \$120 each. What inequalities represent the number of new shoes the team can purchase?

18) From last year, the price of gasoline has increased from \$1.40 per gallon to \$1.75 per gallon. The new price is what percent of the original price?

19) 6 liters of water are poured into an aquarium that's 25 cm long, 5 cm wide, and 60 cm high. How many cm will the water level in the aquarium rise due to this added water? (1 $liter\ of\ water = 1{,}000\ cm^3$)

20) If a box contains red and blue balls in ratio of $2:3$, how many red balls are there if 72 blue balls are in the box?

21) Let r and p be constants. If $x^2 + 6x + r$ factors into $(x+2)(x+p)$, what are the values of r and p respectively?

22) If $2 + \frac{3x}{x-5} = \frac{3}{5-x}$, $x = ?$

23) If $f(x) = 7x - 5$ and $g(x) = 2x^2 - 4x$, then find $\left(\frac{f}{g}\right)(x)$.

24) If $\tan\theta = \frac{5}{12}$ and $\sin\theta > 0$, then $\cos\theta = ?$

25) What is the vertical asymptote of the graph $y = \frac{5x-6}{3x+4}$?

26) A boat sails 80 miles south and then 60 miles east. How far is the boat from its start point?

27) What is the product of all possible values of x in the following equation? $|x - 10| = 4$

28) A number is chosen at random from 1 to 20. Find the probability of not selecting a composite number.

29) The first five terms in a geometric sequence are shown,

$$5, -20, 80, -320, 1280, \cdots$$

Based on this information, what is the equation of the nth term in the sequence, a_n?

30) If $y = 4ab + 3b^3$, what is the value of y, when $a = 2$ and $b = 4$?

31) A line connects the midpoint of AB (point E), with point C in the square $ABCD$. Calculate the area of the resulting trapezoid shape if the square has a side of $4\ cm$.

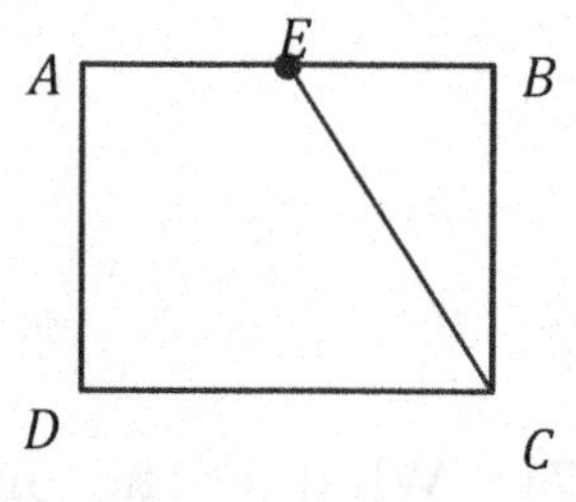

32) Find the value of x in this equation. $log(5x + 2) = log(3x - 1)$

33) Simplify. $-15a(a + b)^2 + 23a(a + b)^2$

34) The population of a colony increases according to the equation $P = 1.7e^{rt}$, where t is the number of months, and r is the rate of growth. What is the equation of r?

35) Find the domain and range of the radical function. $y = 5\sqrt{3x + 9} + 7$

Domain: ___________

Range: ___________

This is the end of Practice Test 6.

ALEKS Mathematics

Practice Test 7

2023 - 2024

Total number of questions: 35

Total time: No time limit

Calculators are permitted for ALEKS Math Test.

1) Solve $m^{\frac{1}{2}}n^{-2}m^{4}n^{\frac{2}{3}}$.

2) Two trains leave the station at the same time, one heading west and the other east. The westbound train travels at 95 miles per hour. The eastbound train travels at 85 miles per hour. How long will it take for the two trains to be 450 miles apart?

3) What is the negative solution to the equation $0 = \frac{2x^2}{5} - 10$?

4) Solve the following system of equations.

$$x + 2y = 10$$
$$6x - 2y = 18$$

5) Find the axis of symmetry of the function $g(x) = -\frac{1}{8}(x-1)^2 - 3$.

6) Quadratic functions f and h are shown below.

$$f(x) = 3x^2 - 6$$
$$h(x) = 3x^2 + c$$

For what value of c will the graph of h be 12 units above the graph of f?

7) A bank is offering 3.5% simple interest on a savings account. If you deposit \$11,000, how much interest will you earn in two years?

8) If the ratio of home fans to visiting fans in a crowd is $3:2$ and all 24,000 seats in a stadium are filled, how many visiting fans are in attendance?

9) If the interior angles of a quadrilateral are in the ratio $2:3:3:4$, what is the measure of the largest angle?

10) If $x = 9$, what is the value of y in the following equation? $2y = \frac{2x^2}{3} + 6$

11) The length of a rectangle is $\frac{5}{4}$ times its width. If the width is 20 cm, what is the perimeter of this rectangle?

12) The perimeter of the trapezoid below is 36 cm. What is its area?

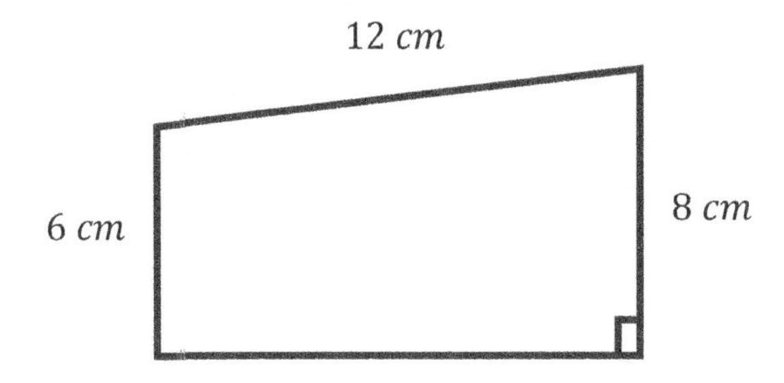

13) An angle is equal to one-ninth of its supplement. What is the measure of that angle?

14) If $\frac{5}{x+1} = \frac{x+1}{x^2-1}$, then $x =$ ____

15) If the function is defined as $f(x) = bx^2 + 15$, b is a constant, and $f(2) = 35$, what is the value of $f(3)$?

16) If $sin\ A = \frac{1}{3}$ in a right triangle and the angle A is an acute angle, then what is $cos\ A$?

17) In the standard (x, y) coordinate system plane, what is the area of the circle with the following equation?

$$(x + 2)^2 + (y - 4)^2 = 25$$

18) Solve the equation: $log_3(x + 20) - log_3(x + 2) = 1$.

19) If the equation $y = (x + 3)(x - 9)$ is graphed in the xy −plane, what is the y −coordinate of the parabola's vertex?

20) If 150% of a number is 75, then what is 80% of that number?

21) Find the center and the radius of a circle with the following equation.
$x^2 + y^2 - 6x + 4y + 4 = 0$

Center: (_,_)

radius = _____

22) The expression $6x^2 + 4x - 10$ can be written in factored form as $(2x - m)(3x + 5)$, where m represents a number. What is the value of m?

23) If $\tan x = \frac{8}{15}$, then $\sin x =$?

24) $(x^7)^{\frac{9}{8}}$ equal to:

25) Create a cubic function that has roots at $x = -1$, 1, and 3.

26) If $x + sin^2 a + cos^2 a = 3$, then $x = ?$

27) If $\sqrt{6x} = \sqrt{y}$, then $x =$

28) The average weight of 18 girls in a class is 57 kg and the average weight of 32 boys in the same class is 65 kg. What is the average weight of all 50 students in that class?

29) How many x –intercepts does the graph of $y = \frac{x-1}{1-x^2}$ have?

30) Sketch the graph of $y = (x + 1)^2 - 2$.

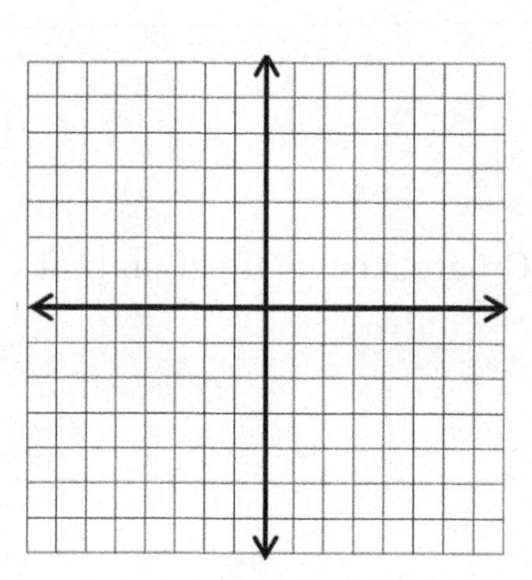

31) What is the value of x in the following equation?

$$\frac{x^2-9}{x+3}+2(x+4)=15$$

32) Evaluate the expression $7x^2-2xy+y^2$, when $x=4$ and $y=9$.

33) $f(x)=ax^2+bx+c$ is a quadratic function where a, b, and c are constant. The value of x of the point of intersection of this quadratic function and the linear function $g(x)=2x-3$ is 2. The vertex of $f(x)$ is at $(-2,5)$. What is the product of a, b, and c?

34) Simplify this rational expression: $\frac{5x}{x+2}\div\frac{x}{3x+6}=$

35) Solve this equation for x: $e^{2x}=12$

This is the end of Practice Test 7.

ALEKS Mathematics

Practice Test 8

2023 - 2024

Total number of questions: 35

Total time: No time limit

Calculators are permitted for ALEKS Math Test.

1) The equation $x^2 = 4x - 3$ has how many distinct real solutions?

2) Evaluate.
$(-8)^3 =$
$5^{-4} =$

3) Point A lies on the line with equation $y - 3 = 2(x + 5)$. If the x-coordinate of A is 8, what is the y-coordinate of A?

4) If $x^2 + 3$ and $x^2 - 3$ are two factors of the polynomial $12x^4 + n$ and n is a constant, what is the value of n?

5) Factor this polynomial completely.
$5v^3 - 3v^2 + 20v - 12$

6) What is the ratio of the minimum value to the maximum value of the following function?

$$f(x) = -3x + 1; -2 \le x \le 3$$

7) The graph of $y = f(x)$ in the xy –plane is shown below. What is the value of $f(0)$?

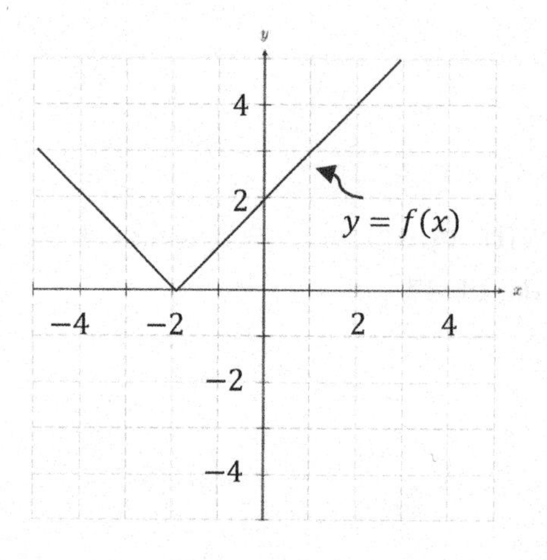

8) If the following equations are true, what is the value of x?

$$a = \sqrt{5}$$
$$4a = \sqrt{4x}$$

9) In a sequence of numbers, $a_4 = 19$, $a_5 = 23$, $a_6 = 27$, $a_7 = 31$, and $a_8 = 35$. Based on this information, which equation can be used to find the nth term in the sequence, a_n?

10) Multiply and write the product in scientific notation:
$(4.09 \times 10^{6}) \times (5.6 \times 10^{-4})$

11) If $x \neq -4$ and $x \neq 5$, simplify $\frac{1}{\frac{1}{x-5}+\frac{1}{x+4}}$.

12) In the xy –plane, if a point with coordinates (a, b) lies in the solution set of the system of inequalities below, what is the maximum possible value of b?

$$y \leq -15x + 3{,}000$$

$$y \leq 5x$$

13) In the following figure, $ABCD$ is a rectangle, and E and F are points on AD and DC, respectively. The area of ΔBED is 16, and the area of ΔBDF is 18. What is the perimeter of the rectangle?

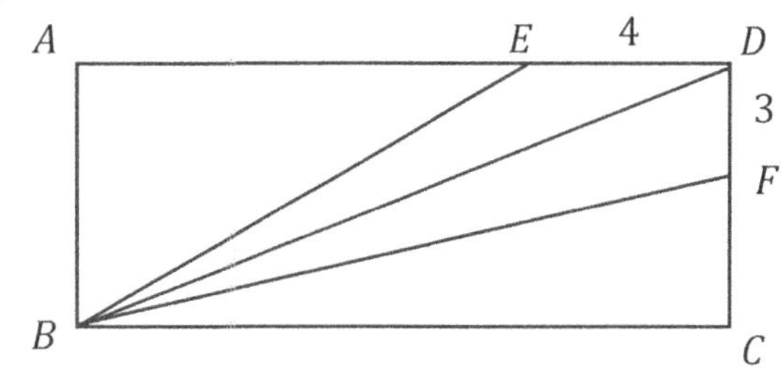

14) What is the value of the y –intercept of the graph $f(x) = 11.5(0.8)^{x}$?

15) The perimeter of a rectangular garden is 38 meters. The length of the garden can be represented by $(x + 6)$ meters, and its width can be represented by $(2x - 2)$ meters. What are the dimensions of this garden in meters?

16) Convert radian measure $\frac{7\pi}{3}$ to degree measure.

17) Find positive and negative coterminal angles to angle $\frac{\pi}{3}$.

18) In the following figure, $ABCD$ is a rectangle. If $a = \sqrt{3}$, and $b = 2a$, find the area of the shaded region. (The shaded region is a trapezoid)

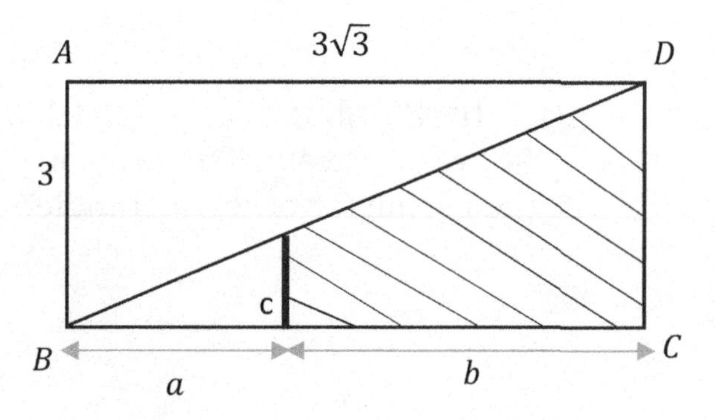

19) The surface area of a cylinder is $150\pi\ cm^2$. If its height is $10\ cm$, what is the radius of the cylinder?

20) A construction company is building a wall. The company can build 25 cm of the wall per minute. After 50 minutes $\frac{3}{4}$ of the wall is completed. How many meters is the wall?

21) The value of $\frac{8+3x}{7} - \frac{4-4x}{7}$ is how much greater than the value of x.

22) The perimeter of a rectangular yard is 72 meters. What is its length if its width is twice its length?

23) If the ratio of $5a$ to $2b$ is $\frac{1}{20}$, what is the ratio of a to b?

24) If θ is an acute angle and $\sin\theta = \frac{4}{5}$ then $\cos\theta = ?$

25) If 60% of x is equal to 30% of 20, then what is the value of $(x+5)^2$?

26) If $f(x) = \frac{10x-3}{6}$ and $f^{-1}(x)$ is the inverse of $f(x)$, what is the value of $f^{-1}(2)$?

27) What is the value of x in the following equation? $log_4(x+2) - log_4(x-2) =$ 1

28) If $4x = \frac{48}{3}$, what is the value of 7^{x-2}?

29) What is the area of the following equilateral triangle if the side $AB = 8\ cm$?

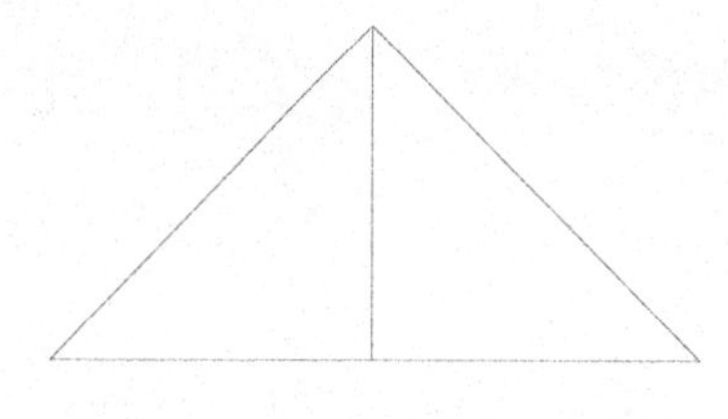

30) What is the sum of all values of n that satisfy $2n^2 + 16n + 24 = 0$?

31) Simplify. $\frac{(4u^{-5}v^3)^{-2}}{6w}$
Write the expression with positive powers.

32) If a parabola with equation $y = ax^2 + 5x + 10$, where a is constant, passes through the point $(2, 12)$, what is the value of a^2?

33) Simplify $\frac{6}{\sqrt{12}-3}$.

34) Simplify and express in the form $a + bi$: $\frac{-6i}{15+i}$

35) Sketch the graph of $2y > 4x^2$.

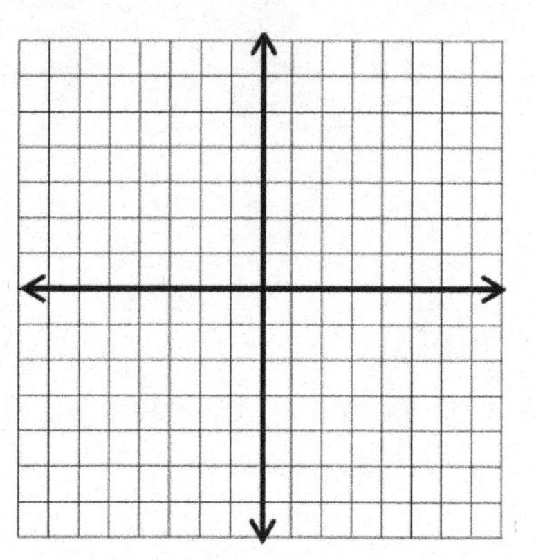

This is the end of Practice Test 8.

ALEKS Mathematics

Practice Test 9

2023 - 2024

Total number of questions: 35

Total time: No time limit

Calculators are permitted for ALEKS Math Test.

1) Factor.
   $12x^8 - 21x^4 + 3x^3$

2) Solve for x.
   $3(7^{2x}) = 11$

3) Two consecutive odd integers have a sum of 28. Find the integers.

4) The equation of a line is given below. $-8x - 2y = 4$
   Find the slope and y-intercept.
   Then use them to graph the line.
   Slope: ___________

   y-intercept: ___________

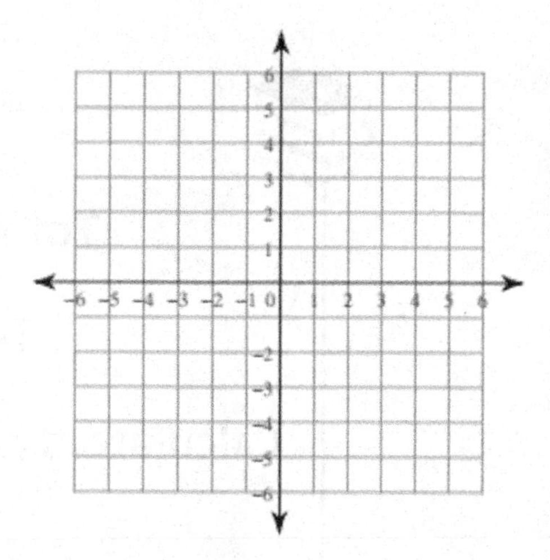

5) What is the vertex of the graph of the quadratic function $g(x) = 3x^2 + 4x + 2$?

6) What is the value of y in the following system of equations?

$$2x + 5y = 11$$
$$4x - 2y = -14$$

7) The first three terms of a geometric sequence are 4, −2, and 1. Find the nth term of the sequence.

8) Write as a single fraction.

$$-4 + \frac{2a - 6y}{12a} - \frac{6a + 4y}{8a}$$

Simplify your answer as much as possible.

9) If $f(x) = 3x^3 + 5x^2 + 2x$ and $g(x) = -2$, what is the value of $f(g(x))$?

10) The average annual energy cost for a particular home is \$4,334. The homeowner plans to spend \$25,000 to install a geothermal heating system. The homeowner estimates that the average annual energy cost will be \$2,712. What inequality can be used to find t, the number of years after installation at which the total energy cost saving will exceed the installation cost?

11) If one angle of a right triangle measures 30°, what is the sine of the other acute angle?

12) In the figure below, line A is parallel to line B. What is the value of angle x?

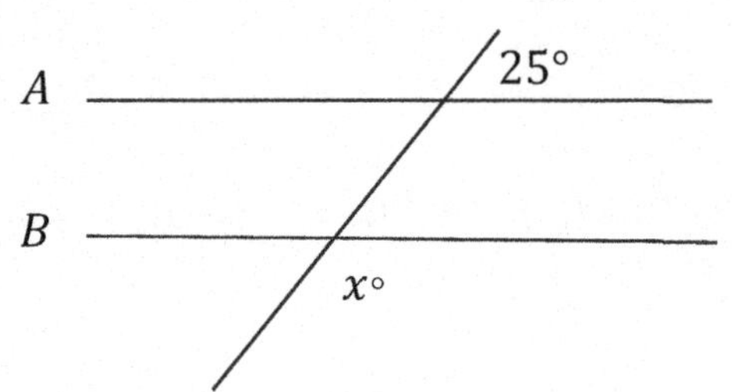

13) An angle is equal to one-fifth of its supplement. What is the measure of that angle?

14) The cost of using a car is \$0.35 per minute. Find the equation that represents the total cost c, in dollars, for h hours of using the car.

15) The average of five consecutive numbers is 40. What is the smallest number?

16) If $sin\, A = \frac{1}{4}$ in a right triangle and angle A is an acute angle, then what is $cos\, A$?

17) In the standard (x, y) coordinate system plane, what is the area of the circle with the following equation?

$$(x+2)^2 + (y-4)^2 = 16$$

18) What is the slope of a line that is perpendicular to the line $6x - 3y = 15$?

19) Solve for x.

$$e^2 3^{1-x} = \frac{10}{5^{2x}}$$

20) If 140% of a number is 70, then what is 85% of that number?

21) If the cotangent of an angle β is 1, then the tangent of angle β is ...

22) Simplify this expression: $\sqrt{\frac{x^2}{2}+\frac{x^2}{16}}$.

23) If $tan\, x = \frac{8}{15}$, then $sin\, x =$

24) What is the range of the function shown below?

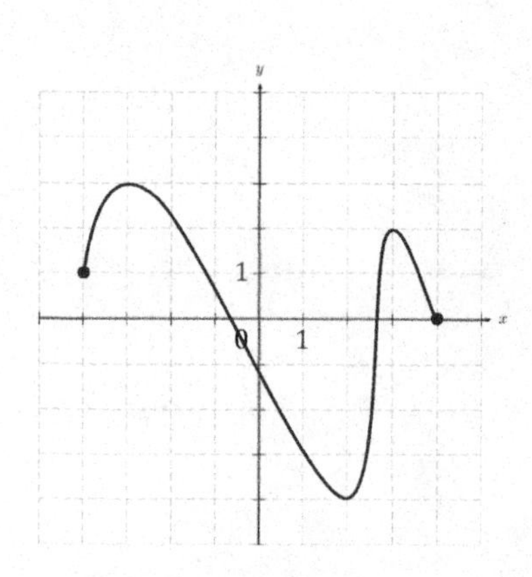

25) What are the zeroes of the function $f(x) = x^3 + 6x^2 + 8x$?

26) Simplify as much as possible: $\frac{6}{x^3} + \frac{5x-3}{x^4}$.

27) In the following figure, point Q lies on line n, what is the value of y if $x = 35$?

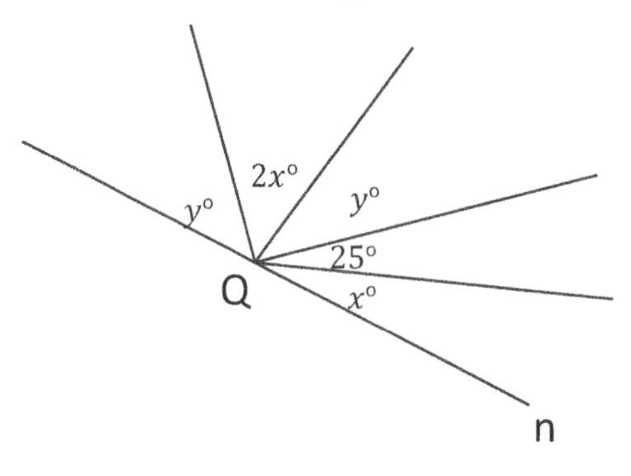

28) Let A, B, and C be three sets as follows:

$$A = \{2,4,6,8,10,12\}$$
$$B = \{4,8,12,16\}$$
$$C = \{1,3,4,7,11,18\}$$

Find $(B \cup C) \cap A$.

29) The diagonal of a rectangle is 10 inches long and the height of the rectangle is 8 inches. What is the perimeter of the rectangle?

30) Sketch the graph of $y + 3 = 2(x - 1)$.

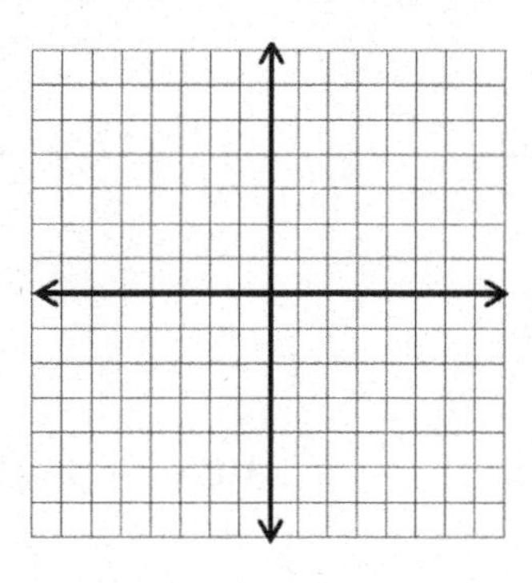

31) Perform the indicated operation and write the result in standard form:

$$(-8 + 11i)(-6 - 16i)$$

32) If $a - b = 2$ and $a^3 + b^3 = 32$, find ab.

33) What is the least common denominator for $\frac{3x}{x^2-36}$ and $\frac{4}{2x-12}$?

34) Find AC in the following triangle. Round your answer to the nearest tenth.

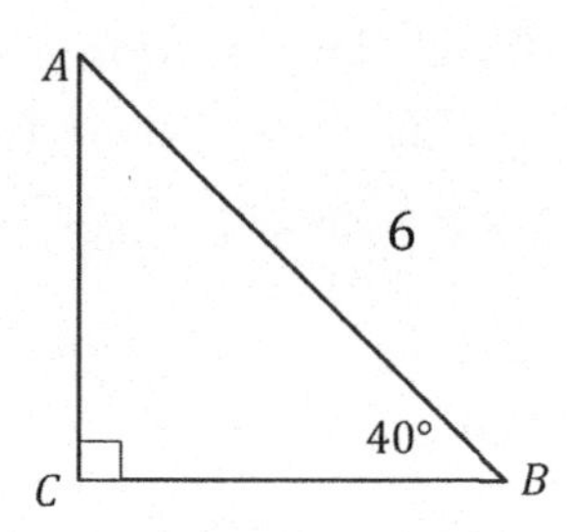

35) Given $f(x^3) = 2x - 5$, for all values of x, what is the value of $f(8)$?

This is the end of Practice Test 9.

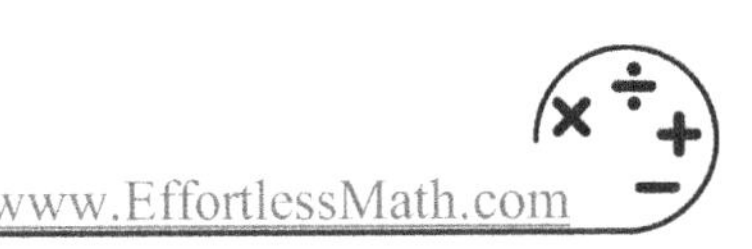

ALEKS Mathematics

Practice Test 10

2023 - 2024

Total number of questions: 35

Total time: No time limit

Calculators are permitted for ALEKS Math Test.

1) What is the equation of the following graph?

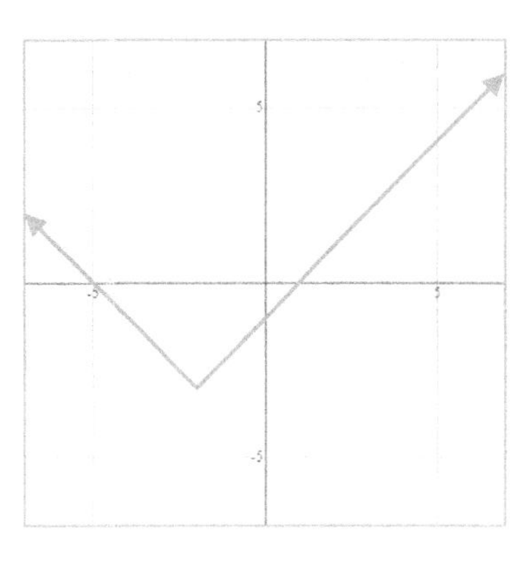

2) Multiply.
$4x^4u^5 \times 6u^2 \times 7x$
Simplify your answer as much as possible.

3) Solve for n, where $n \in \mathbb{N}$: $\frac{(n+2)!}{(n-1)!} = 24$.

4) Factor by grouping (sometime called the ac-method).
$4x^2 - 4x - 15$
First, choose a form with appropriate signs.
Then, fill in the blanks with numbers to be used for grouping.
Finally, show the factorization.

5) Two third of 15 is equal to $\frac{2}{5}$ of what number?

6) For what value of x is $|x - 3| + 3$ equal to 0?

7) $5 + 8 \times (-3) - [4 + 22 \times 5] \div 6 = ?$

8) The score of Emma was half that as that of Ava and the score of Mia was twice that of Ava. If the score of Mia was 40, what is the score of Emma?

9) Find $tan\ \frac{2\pi}{3}$.

10) What is the value of x in this equation?

$$4\sqrt{2x + 6} = 24$$

11) Write in terms of $log(x)$, $log(y)$:

$$log\frac{\sqrt[3]{y}}{10yx^2}$$

12) If the center of a circle is at the point $(-4, 2)$ and its circumference equals to $2\,\pi$, what is the standard form equation of the circle?

13) Find the value of x in the following diagram.

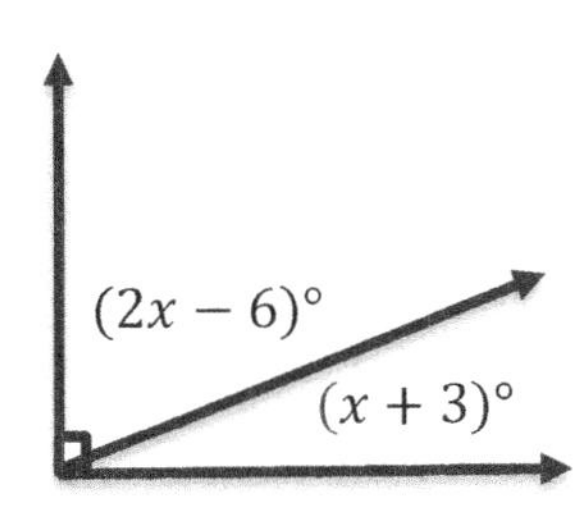

14) When a number is subtracted from 28 and the difference is divided by that number, the result is 3. What is the value of the number?

15) Find a positive and a negative Coterminal angle to angle 115°.

Positive coterminal angle of 115°: _____

Negative coterminal angle of 115°: _____

16) Solve inequality and graph it. $8x - 4 \geq -2x + 16$

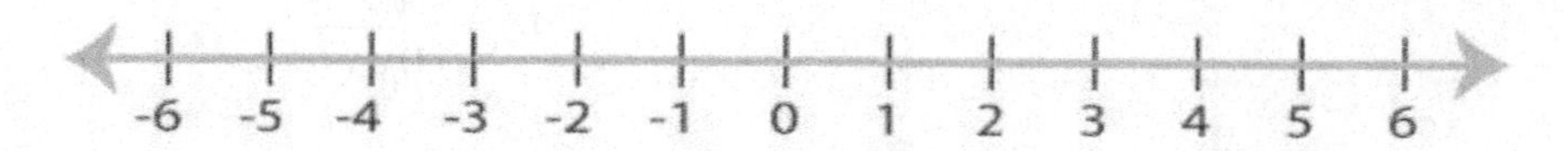

17) Find the zeros of the following function: $h(x) = x^5 - x^2$

18) Solve for x: $3(12x + 4) = -2(x + 2)$

19) What is the distance between the points $(1, 3)$ and $(-2, 7)$ on the coordinate plane?

20) The interval solution to the inequality $\frac{6x+12}{x-8} > 0$ is:

Solution: __________

Interval Notation: __________

21) The average of five numbers is 26. If a sixth number 56 is added, then, what is the new average?

22) If $a + 2b = c$ and $abc = 12$. What is the value of $a^3 + 8b^3 - c^3$?

23) Write the expanded form of $(1 - c)(2a - b + 1)^2$.

24) 5 less than twice a positive integer is 73. What is the integer?

25) Find the solutions of the following equation.

$$x^2 + 2x - 5 = 0$$

26) Sketch the graph of linear inequality: $y \leq -2x - 2$

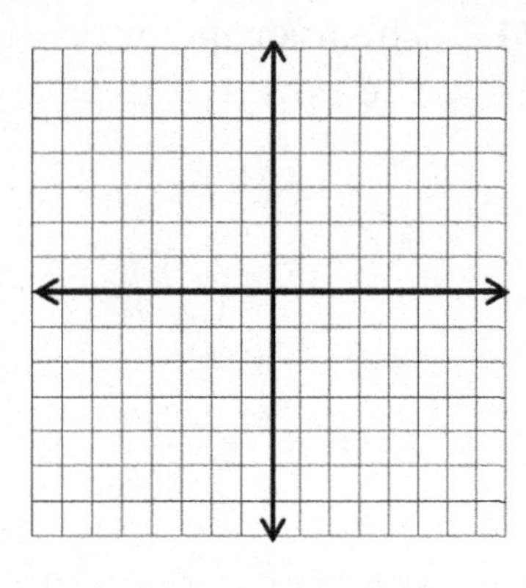

27) What is the domain of the function $f(x) = \frac{x^3-1}{x^2-4x+3}$?

28) If $\log_2 x = 5$, then $x =$?

29) What is the value of $\frac{10b}{c}$ when $\frac{c}{b} = 5$?

30) What is the equivalent temperature of $140°F$ in Celsius?

$$C = \frac{5}{9}(F - 32)$$

31) The first five terms in row 30 of Pascal's triangle are 1, 30, 435, 4,060, and 27,405. Find the first five terms in row 31.

32) Find the equation of the horizontal asymptote of the function $f(x) = \frac{x+3}{x^2+1}$.

33) Write Find the value of x in the following triangle. (Round your answer to the whole number)

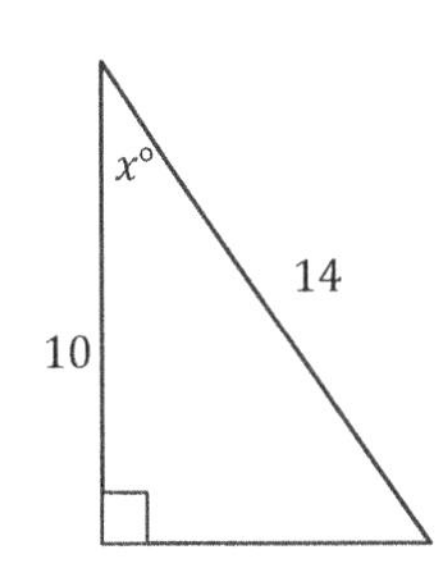

34) Convert 135 degrees to radian.

35) Solve and write the answer in scientific notation: $(8.5 \times 10^8) - (4.2 \times 10^7)$

This is the end of Practice Test 10.

ALEKS Mathematics Practice Tests Answers and Explanations

Now, it's time to review your results to see where you went wrong and what areas you need to improve!

ALEKS Mathematics Practice Test 1

1) The answer is $1\frac{5}{8}$.

$\frac{48}{5}=\frac{6}{x-1}$. Cross multiply: $48(x-1)=6(5)$ apply multiplicative inverse property; divide both sides of the equation by 48. Then, simplify the fraction:

$x-1=\frac{30}{48}\rightarrow x=\frac{30}{48}+1=\frac{78}{48}\rightarrow x=1\frac{30}{48}=1\frac{5}{8}$

2) The answer is $x=-2$ *and* $y=3$.

$\begin{cases} x+4y=10 \\ 5x+10y=20 \end{cases}$ → Multiply the top equation by -5 then, $\begin{cases} -5x-20y=-50 \\ 5x+10y=20 \end{cases}$

Add two equations: $-10y=-30\rightarrow y=3$, plug in the value of y into the first equation

$x+4y=10\rightarrow x+4(3)=10\rightarrow x+12=10$

Subtract 12 from both sides of the equation. Then: $x+12=10\rightarrow x=-2$

3) The answer is $2.

Let's use the variable c to represent the cost of each notebook. Since we know that the cost of 8 notebooks is $16, we can write the equation: $8c=16$

To solve for c, we can divide both sides of the equation by 8: $c=2$

Therefore, the cost of each notebook at this store is $2.

4) The value of x is 33.

Since $N=6$, substitute 6 for N in the equation $\frac{x-3}{5}=N$, which gives $\frac{x-3}{5}=6$. Multiplying both sides of $\frac{x-3}{5}=6$ by 5 gives $x-3=30$ and then adding 3 to both sides of $x-3=30$ then,

$x=33$.

5) The answer is 28%.

John has not completed 7 out of 25 problems in his physics homework.

To find the percentage of problems John has not completed, we can use the formula:

$Percentage=\left(\frac{number\ of\ problems\ not\ completed}{total\ number\ of\ problems}\right)\times 100\%$

So, in this case, the percentage of problems John has not completed is $\left(\frac{7}{25}\right)\times 100\%=28\%$.

6) The answer is 20.

Since $(x+y)^2 = x^2 + 2xy + y^2$, then:

$2xy = (x+y)^2 - x^2 - y^2 \rightarrow xy = \frac{(x+y)^2-(x^2+y^2)}{2}$.

Substitute $x + y = 7$ and $x^2 + y^2 = 9$. So,

$xy = \frac{(7)^2-(9)}{2} = 20.$

7) The answer is $4 - 7i$.

We know that: $i = \sqrt{-1} \Rightarrow i^2 = -1$

$3i(2i-3) + (1-5i)(2i) = (3i \times 2i - 3i \times 3) + (1 \times 2i - 5i \times 2i)$

$= (6i^2 - 9i) + (2i - 10i^2)$

$= 6(-1) - 9i + 2i - 10(-1)$

$= 4 - 7i$

8) The answer is $\sum_{k=0}^{\infty} \frac{(-1)^{n+1}}{3n+1} r^n$.

Considering that the explicit formula of the series in the problem is $\frac{(-1)^{n+1}}{3n+1} r^n$, it is enough to determine the lower and upper limits of the series. Let $\frac{(-1)^{n+1}}{3n+1} r^n = -1$ to set the lower limit. Then, $n = 0$. Since the series is infinite, the upper limit is also infinite.

$-1 + \frac{r}{4} - \frac{r^2}{7} + \frac{r^3}{10} - \frac{r^4}{13} + \cdots + \frac{(-1)^{n+1}}{3n+1} r^n + \cdots = \sum_{k=0}^{\infty} \frac{(-1)^{n+1}}{3n+1} r^n.$

9) The answer is 2.7.

Applying the zero product property the solutions to $(3-x)(x+0.3) = 0$ are $(3-x) = 0 \rightarrow x = 3$ and $(x+0.3) = 0 \rightarrow x = -0.3$.

$3 + (-0.3) = 2.7$

10) The answer is -1.

Rewrite the equation of the function $h(x)$ as in the vertex form:

$h(x) = x^2 + 2x - 1 \rightarrow h(x) = (x^2 + 2x + 1 - 1) - 1 = (x+1)^2 - 2$

Since the function $g(x)$ is 2 units higher than $h(x)$, it follows $g(x) = h(x) + 2$. So,

$g(x) = ((x+1)^2 - 2) + 2 \rightarrow g(x) = (x+1)^2$

Therefore, by comparing the given function for g and the obtained function, we find the value of c: $(x-c)^2=(x+1)^2 \rightarrow -c=1 \rightarrow c=-1$.

11) The answer is $g(x)=-2x+1$

The points are $(1,-1)$, $(2,-3)$, and $(3,-5)$ then: $g(x)=-2x+1$

12) The value of $a+b$ is 4.

By using: $f(1)=1 \rightarrow f(1)=a2^{b\times 1-1}-1=1 \rightarrow a2^0-1=1 \rightarrow a-1=1 \rightarrow a=2$,

and $f(2)=15 \rightarrow f(2)=2\times 2^{b\times 2-1}-1=15 \rightarrow 2\times 2^{2b-1}-1=15 \rightarrow 2\times 2^{2b-1}=16$,

$2^{2b-1}=8 \rightarrow 2b-1=3\ (since\ 2^3=8) \rightarrow 2b=4 \rightarrow b=2$

Thus, $a=2$ and $b=2$.

The value of $a+b$ is:

$2+2=4$

13) The answer is -4.

To find the y −intercept of a line from its equation, put the equation in slope-intercept form:

$x-3y=12,\ -3y=-x+12,\qquad 3y=x-12,\qquad y=\frac{1}{3}x-4$

$y=\frac{1}{3}(0)-4 \rightarrow y=-4$

Thus, the y −intercept of the line is -4.

14) The answer is 192.

Rewrite the exponential form: $\sqrt[3]{9}\times 2^4\left(\frac{128}{3^{-1}\times\sqrt[4]{16}}\right)^{\frac{1}{3}}=3^{\frac{2}{3}}\times 2^4\left(\frac{2^7}{3^{-1}\times 2^{\frac{4}{4}}}\right)^{\frac{1}{3}}=3^{\frac{2}{3}}\times 2^4\left(\frac{2^7}{3^{-1}\times 2}\right)^{\frac{1}{3}}$.

Use the rule: $\frac{v^m}{v^n}=v^{m-n}$ and $\frac{1}{a^b}=a^{-b}$. Therefore,

$3^{\frac{2}{3}}\times 2^4\left(\frac{2^7}{3^{-1}\times 2}\right)^{\frac{1}{3}}=3^{\frac{2}{3}}\times 2^4(2^{7-1}\times 3)^{\frac{1}{3}}=3^{\frac{2}{3}}2^4(2^6\times 3)^{\frac{1}{3}}$.

We know that: $(u^m)^n=u^{mn}$. Then:

$3^{\frac{2}{3}}\times 2^4(2^6\times 3)^{\frac{1}{3}}=3^{\frac{2}{3}}\times 2^4(2^6)^{\frac{1}{3}}\times 3^{\frac{1}{3}}=3^{\frac{2}{3}}\times 2^4\times 2^{6\times\frac{1}{3}}\times 3^{\frac{1}{3}}=3^{\frac{2}{3}}\times 2^4\times 2^2\times 3^{\frac{1}{3}}$.

By using the rule: $a^m\times a^n=a^{m+n}$. Finally:

$3^{\frac{2}{3}}\times 2^4\times 2^2\times 3^{\frac{1}{3}}=\left(3^{\frac{2}{3}}\times 3^{\frac{1}{3}}\right)\times(2^4\times 2^2)=3^{\frac{2}{3}+\frac{1}{3}}\times 2^{4+2}=3\times 2^6=192$.

15) The answer is 5%.

Use the percent increase expression to find the answer:

$$Percent\ increase = \frac{new\ price - original\ price}{original\ price} \times 100\% = \frac{5.67 - 5.40}{5.40} \times 100\% = 5\%$$

16) The answer is $3,500$.

To find the number of cases of chocolate bars that the company needs to sell to achieve equal costs and revenue, we need to determine the x –value at which the two lines in the graph intersect.

The fixed cost of $\$21,000$ represents the y –intercept of the cost line, and the slope of the cost line is the variable cost of $\$4$ per case. Therefore, the equation of the cost line is:

Cost $= 4x + 21,000$.

The revenue line represents the income earned from selling x cases of chocolate bars, and since the selling price is $\$10$ per case, the slope of the revenue line is $\$10$ per case. Therefore, the equation of the revenue line is: Revenue $= 10x$.

To find the point of intersection, we can set the two equations equal to each other: $4x + 21,000 = 10x$.

Subtracting $4x$ from both sides gives: $21,000 = 6x$.

Dividing both sides by 6 gives: $x = 3,500$.

Therefore, the company needs to sell $3,500$ cases of chocolate bars to achieve equal costs and revenue.

17) The answer is 5 hours and 20 minutes.

To find the number of hours and minutes that the car has been traveling if the distance between the two cities is 500 kilometers, we can start by setting the equation $k = 60t + 180$ equal to 500 and solving for t:

$k = 500 \rightarrow 500 = 60t + 180 \rightarrow 60t = 320 \rightarrow t = \frac{16}{3}$.

So, the car has been traveling for $\frac{16}{3}$ hours. To find the number of minutes, we convert the time passed into minutes: $\frac{16}{3} \times 60 = 320$.

320 minutes includes 5 hours (300 minutes) and 20 minutes.

So, the answer is 5 hours and 20 minutes.

18) The answer is $-\frac{25}{4}$.

$\frac{|1-2(4+1)|}{|-(3-15)|}-7=\frac{|1-2(5)|}{|-(-12)|}-7=\frac{|1-10|}{|12|}-7=\frac{|-9|}{|12|}-7=\frac{9}{12}-7=\frac{3}{4}-7=\frac{3-28}{4}=-\frac{25}{4}$.

19) The answer is $\frac{243e^6(e^{12}-1)}{8}$.

To find the function that models population growth, we need to find the population function like $g(t)$. For this purpose, use the formula for population growth with $g(0)=243$ and $r=\frac{3}{2}$. Then,

$g(t)=243e^{\frac{3}{2}t}$.

To calculate the average rate of change for the population growth $g(t)=243e^{\frac{3}{2}t}$, put $t_1=4$ and $t_2=12$ with the corresponding value $g(4)=243e^6$ and $g(12)=243e^{18}$.

Use this formula $\frac{f(b)-f(a)}{b-a}$ as follows: $\frac{f(12)-f(4)}{12-4}=\frac{243e^{18}-243e^6}{12-4}=\frac{243e^6(e^{12}-1)}{8}$.

20) The answer is 682.

Using the finite geometric series formula: $S_n=\sum_{i=1}^{n} ar^{i-1}=a_1\left(\frac{1-r^n}{1-r}\right)$. Substitute $a_1=2$ and $r=4$ in the previous formula. Now, $S_5=2\left(\frac{1-4^5}{1-4}\right)=2\left(\frac{1-1024}{1-4}\right)=2\left(\frac{-1023}{-3}\right)\rightarrow S_5=682$.

21) The answer is 5.

Let x be the number of adult tickets and y be the number of student tickets. Then:

$x+y=12$
$12.50x+7.50y=125$

Use the elimination method to solve this system of equations. Multiply the first equation by -7.5 and add it to the second equation: $-7.5(x+y=12)\rightarrow -7.5x-7.5y=-90$.

$12.50x+7.50y+(-7.5x-7.5y)=125+(-90)\rightarrow 5x=35\rightarrow x=7$

Next, substitute $x=7$ in one of the equations. So,

$x=7\rightarrow (7)+y=12\rightarrow y=5$

There are 7 adult tickets and 5 student tickets.

22) The answer is 25.

Write the ratios in fraction form and solve for x: $\frac{x}{40}=\frac{20}{32}$.

Cross multiply: $32x = 800$. Then apply the multiplicative inverse property;

Divide both sides by 32: $x = \frac{800}{32} = 25$.

23) The answer is $800x + 250y \leq 10,000$.

To solve this problem, we can use the given information to write an inequality in terms of the number of laptops and printers the company can purchase while staying within its budget of $\$10{,}000$.

Let x be the number of laptops and y be the number of printers. Each laptop costs $\$800$, so the cost of x laptops is $800x$. Similarly, each printer costs $\$250$, so the cost of y printers is $250y$. The total cost of the laptops and printers must be less than or equal to the budget of $\$10{,}000$, so we can write: $800x + 250y \leq 10{,}000$.

24) The answer is 54.

The amount of money that Jack earns for one hour: $\frac{\$616}{44} = \$14$.

The number of additional hours that he needs to works to make enough money is: $\frac{\$826-\$616}{1.5\times\$14} = 10$.

The total number of hours is: $44 + 10 = 54$.

25) The answer is 820.

To find any term in an arithmetic sequence use this formula: $a_n = a_1 + d(n-1)$.

We have $a_2 = 7$ and $a_8 = 31$.

Therefore, $a_2 = a_1 + d(2-1) \rightarrow a_1 + d = 7$ and $a_8 = a_1 + d(8-1) \rightarrow a_1 + 7d = 31$.

Solve the following equation system:

$$\begin{cases} a_1 + d = 7 \\ a_1 + 7d = 31 \end{cases}$$

Subtracting the second equation from the first equation, we are left with:

$(a_1 + 7d) - (a_1 + d) = 31 - 7$.

Simplify, $a_1 + 7d - a_1 - d = 31 - 7 \rightarrow 6d = 24 \rightarrow d = 4$. So, by substituting $d = 4$ in the first equation: $a_1 + d = 7 \rightarrow a_1 + 4 = 7 \rightarrow a_1 = 3$. Then, use this formula: $S_n = \frac{n}{2}(a_1 + a_n)$.

$S_n = \frac{n}{2}(a_1 + a_n) \rightarrow S_n = \frac{n}{2}\big(a_1 + a_1 + d(n-1)\big) \rightarrow S_n = \frac{n}{2}\big(2a_1 + d(n-1)\big)$.

Now, $S_{20} = \frac{20}{2}(2(3) + 4(20 - 1)) \rightarrow S_{20} = 10(6 + 76) \rightarrow S_{20} = 820.$

26) The answer is 40.

If 20 students are in both the theater group and football team, then 20 students are only in the theater group and 40 are only in the football team. The total in the three categories is $20 + 20 + 40 = 80$. This means that out of all 120 students, only 80 are active in these two categories. That leaves 40 students that are not in either.

27) The answer is 8.

Since $A \cup B$ represents the union of A and B. Rewrite the members of the two sets A and B as follows:

$A = \{4,6,8,10,12,14\}$, and $B = \{14,21,28\}$.

The numbers that A and B have in union are $\{4,6,8,10,12,14,21,28\}$ which is 8 elements.

28) The answer is 4.

Use the simple interest formula: $I = prt\ (I = interest\ , p = principal, r = rate, t = time)$

$I = prt \rightarrow 600 = (3{,}000)(0.05)(t) \rightarrow 600 = 150t \rightarrow t = 4$

29) The answer is $\frac{1}{2}x + 42$.

The difference between the two terms in the sequence is 3. $(\frac{1}{2}x + 3 - \frac{1}{2}x = 3)$

To find any term in an arithmetic sequence use this formula: $a_n = a_1 + (n - 1)d$

a_1 = the first term, d = the common difference between terms, n = number of items

Then, the 15th term is: $a_{15} = \frac{1}{2}x + ((15 - 1) \times 3) = \frac{1}{2}x + 42.$

30) The answer is 56.

To solve this problem, we can use algebra. Let's use b to represent the number of boys in the classroom. Then the number of girls (g) is $b + 15$. We know that the total number of students is 97, so we can write: $b + g = 97 \rightarrow b + (b + 15) = 97.$

Simplifying the left side of the equation, we get: $b + b + 15 = 97 \rightarrow 2b + 15 = 97.$

Subtracting 15 from both sides, we get: $2b = 82$. Dividing both sides by 2, we get: $b = 41.$

So, there are 41 boys in the classroom, and the number of girls is $41 + 15 = 56$. Therefore, the total number of girls in the classroom is 56.

31) The values of N and A are 2 and -2, respectively.

Since the remainder of an arbitrary polynomial as $p(x)$ divided by $x - r$ is $p(r)$. By replacing the contents of the problem in the polynomial equation, the following system of equations is obtained, as follows:

$$\begin{matrix} x+2 \rightarrow P(-2) = 2 \\ x+1 \rightarrow P(-1) = -1 \end{matrix} \rightarrow \begin{cases} (-2)^N + A = 2 \\ (-1)^N + A = -1 \end{cases}.$$

Subtract the second equation from the first equation:

$(-2)^N + A - ((-1)^N + A) = 2 - (-1) \rightarrow (-2)^N - (-1)^N = 3 \rightarrow N = 2.$

Substitute $N = 2$ in the first equation, then: $(-2)^2 + A = 2 \rightarrow 4 + A = 2 \rightarrow A = -2.$

32) The answer is 2.

Substituting $x = 4$ into the given function, we get $g(4) = 3(4) - 10 = 2$.

33) The formula representing the nth term is $a_n = 6\left(\frac{3}{2}\right)^{n-1}$.

Considering the contents of the question, the starting term is 6. To find the common ratio, divide a_2 by a_1: $r = \frac{a_2}{a_1} = \frac{9}{6} = \frac{3}{2}$. Now, use the geometric sequence formula: $a_n = a_1 r^{(n-1)}$. Then:

$$a_n = 6\left(\frac{3}{2}\right)^{(n-1)}.$$

34) The answer is 126.

Calculate ${}_9C_4$: ${}_9C_4 = \frac{9!}{4!(9-4)!} = \frac{9!}{4!(5)!} = \frac{9\times8\times7\times6\times5!}{4!5!} = \frac{9\times8\times7\times6}{4\times3\times2} = 126.$

35) The answer is 3.

First expand parentheses $x(x^2 - 1)^2 - 8x^3 = -8x \rightarrow x(x^4 - 2x^2 + 1) - 8x^3 = -8x$.

Adding $8x$ to both sides gives:

$x(x^4 - 2x^2 + 1) - 8x^3 + 8x = 0 \rightarrow (x^5 - 2x^3 + x) - 8x^3 + 8x = 0 \rightarrow x^5 - 10x^3 + 9x = 0.$

$x(x^4 - 10x^2 + 9) = 0$. The expression in parentheses is a quadratic equation in x^2 that can be factored: $x(x^2 - 1)(x^2 - 9) = 0$. These further factors as $x(x - 1)(x + 1)(x - 3)(x + 3) = 0$.

The solutions for x are $x = 0$, $x = -1$, $x = 1$, $x = 3$, $x = -3$. So, the biggest value of x in the equation is 3.

ALEKS Mathematics Practice Test 2

1) The answer is $9x^2 - 30x + 25$.

Use the FOIL (First-Out-In-Last) method to simplify the expression:

$(3x-5)^2 = (3x-5)(3x-5) = 9x^2 - 15x - 15x + 25 = 9x^2 - 30x + 25.$

2) The answer is $-2i$.

Use the FOIL (First-Out-In-Last) method: $-(2-i)(i-2) = -2i + 4 + i^2 - 2i = -4i + i^2 + 4.$

Combine like terms: $(2i-3) - 4i + i^2 + 4 = -2i + i^2 + 1 = -2i + (-1) + 1 = -2i$

3) The answer is $\frac{8x^2-5x-8}{(x-3)(3x-2)}$.

Find a common denominator and simplify: $\frac{3x-2}{x-3} - \frac{x-4}{3x-2} = \frac{(3x-2)^2-(x-4)(x-3)}{(x-3)(3x-2)} =$

$\frac{(3x)^2-2(3x)(2)-2^2+(-x+4)(x-3)}{(x-3)(3x-2)} = \frac{9x^2-12x+4-x^2+3x+4x-12}{(x-3)(3x-2)} = \frac{8x^2-5x-8}{(x-3)(3x-2)}$

4) The answer is $x = \frac{5}{2}y + 5$.

Solve for x: $2x - 5y = 10 \rightarrow x - \frac{5}{2}y = 5 \rightarrow x = \frac{5}{2}y + 5$

5) The answer is 21.

$x = \frac{1}{3}$ and $y = \frac{9}{21}$, substitute the values of x and y in the expression and simplify:

$\frac{1}{x} \div \frac{y}{3} \rightarrow \frac{1}{\frac{1}{3}} \div \frac{\frac{9}{21}}{3} \rightarrow \frac{1}{\frac{1}{3}} = 3$ and $\frac{\frac{9}{21}}{3} = \frac{9}{63} = \frac{1}{7}$. Then: $\frac{1}{\frac{1}{3}} \div \frac{\frac{9}{21}}{3} = 3 \div \frac{1}{7} = 3 \times 7 = 21$

6) The value of n is 3 and $\frac{15}{2}$.

First, find the constant term and the coefficient of x, by using the binomial theorem for $\left(\frac{1}{2}x+1\right)^n$ as follows: $\left(\frac{1}{2}x+1\right)^n = \sum_{i=0}^{n}\binom{n}{i}\left(\frac{1}{2}x\right)^{n-i}$. So, expand the series:

$$\sum_{i=0}^{n}\binom{n}{i}\left(\frac{1}{2}x\right)^i = \binom{n}{0}\left(\frac{1}{2}x\right)^0 + \binom{n}{1}\left(\frac{1}{2}x\right)^1 + \binom{n}{2}\left(\frac{1}{2}x\right)^2 + \cdots = 1 + \frac{n}{2}x + \cdots.$$

And expand $(n-x)^2 = n^2 - 2n + x^2$. Now, multiply by $n^2 - 2n + x^2$:

$$(n^2 - 2nx + x^2)\left(1 + \frac{n}{2}x + \cdots\right) = n^2 + \frac{n^3}{2}x - 2nx + \cdots.$$

Simplify, $(n-x)^2\left(\frac{1}{2}x+1\right)^n = n^2 + n\left(\frac{n^2}{2} - 2\right)x + \cdots.$

Match the obtained expression with the right side of the problem:

$$n^2 + n\left(\frac{n^2}{2} - 2\right)x + \cdots = 9 + \frac{15}{2}x + \cdots.$$

Therefore, solve these equations: $n^2 = 9 \rightarrow n = 3$ and $n\left(\frac{n^2}{2} - 2\right) = \frac{15}{2}$.

7) The answer is $1 \le x < 6$.

Solve for x. $-4 \le 4x - 8 < 16$, add 8 to all sides $-4 + 8 < 4x - 8 + 8 < 16 + 8 \Rightarrow$

$4 < 4x < 24$. Divide all sides by 4: $1 \le x < 6$.

8) The answer is 24.

$2x - 4 = 14 \rightarrow 2x = 18 \rightarrow x = 9$. Now, find the value of $3x - 3$.

$3x - 3 = 3(9) - 3 = 24$

9) The answer is $\frac{17+7i}{26}$.

To perform the division $\frac{3+2i}{5+i}$, multiply the numerator and denominator of $\frac{3+2i}{5+i}$ by the conjugate of the denominator, $5 - i$. This gives $\frac{(3+2i)(5-i)}{(5+i)(5-i)} = \frac{15-3i+10i-2i^2}{5^2-i^2}$. Since $i^2 = -1$, this can be simplified to $\frac{15-3i+10i+2}{25+1} = \frac{17+7i}{26}$

10) The answer is 8.

The first term is 5, and the common ratio is 2. Therefore, $a_1 = 5$ and $r = 2$. The finite geometric series formula is $S_n = \sum_{i=1}^{n} ar^{i-1} = a_1\left(\frac{r^n-1}{r-1}\right)$. So, $1275 = 5\left(\frac{2^n-1}{2-1}\right)$. Solve the obtained equation:

$1275 = 5\left(\frac{2^n-1}{2-1}\right) \rightarrow 1275 = 5(2^n - 1) \rightarrow 2^n = 256 \rightarrow 2^n = 2^8 \rightarrow n = 8$.

11) The value of x is 3 and 4.

Use the combination $C(x,3) = \frac{x!}{3!(x-3)!}$, and the permutation definition $P(x,3) = \frac{x!}{3!}$. Set the two expressions equal. Therefore, $\frac{x!}{3(x-3)!} = \frac{x!}{3!}$. Simplifying,

$\frac{x!}{3(x-3)!} = \frac{x!}{3!} \rightarrow (x - 3)! = 1 \rightarrow (x - 3)! = 1! \rightarrow x - 3 = 1 \rightarrow x = 4$.

Or,

$(x - 3)! = 1 \rightarrow (x - 3)! = 0! \rightarrow x - 3 = 0 \rightarrow x = 3$.

Now, evaluate the obtained values for x in $C(x,3) = P(x,3)$. Therefore,

For $x = 4$: $C(4,3) = P(4,3) \rightarrow \frac{4!}{3!(4-3)!} = \frac{4!}{3!} \rightarrow \frac{24}{6\times(1)!} = \frac{24}{6} \rightarrow 1 = 1$. This is true.

For $x = 3$: $C(3,3) = P(3,3) \rightarrow \frac{3!}{3!(3-3)!} = \frac{3!}{3!} \rightarrow \frac{1}{1\times(0)!} = 1$. This is true.

Therefore, the problem has two solutions: 4 and 3.

12) The answer is 37.

$x^2 = 121 \rightarrow x = 11$ (Positive value) or $x = -11$ (Negative value).

Since x is positive, then $f(121) = f(11^2) = 3(11) + 4 = 33 + 4 = 37$

13) The answer is 2.

The line k is parallel to the line $y = \frac{3}{4}x + 3$, so they have the same slopes. The slope of the line $y = \frac{3}{4}x + 3$ is $\frac{3}{4}$. Therefore, the slope of the line k is also equal to $\frac{3}{4}$.

So, the equation of line k in the slope-intercept form is $y = \frac{3}{4}x - 7$.

By placing the point m on the line k, the value of b will be equal to:

$b = \frac{3}{4}(12) - 7 \rightarrow b = 9 - 7 \rightarrow b = 2$.

14) The answer is 11.5.

$3a = 4b \rightarrow b = \frac{3a}{4}$ and $3a = 5c \rightarrow c = \frac{3a}{5}$.

$$a + 2b + 15c = a + \left(2 \times \frac{3a}{4}\right) + \left(15 \times \frac{3a}{5}\right) = a + 1.5a + 9a = 11.5a$$

The value of $a + 2b + 15c$ is 11.5 times the value of a.

15) The answer is $w = 30A - 4$.

The current formula is $A = \frac{4+w}{30}$. Multiplying each side of the equation by 30 gives $30A = 4 + w$.

Subtracting 4 from each side of $30A = 4 + w$ gives $w = 30A - 4$.

16) The value of k is 17.

Number 125 can be written in exponential form $h^{\frac{k}{6}}$, where h and k are positive integers, as follows: $5^{\frac{18}{6}}$, $25^{\frac{9}{6}}$, $125^{\frac{6}{6}}$, $(125)^{\frac{3}{6}}$, $(5^9)^{\frac{2}{6}}$, $(5^{18})^{\frac{1}{6}}$. Hence, if $h^{\frac{k}{6}} = 125$ where h and k are positive integers, then $\frac{k}{6}$ can be $3, \frac{3}{2}, 1, \frac{1}{2}, \frac{1}{3}$, or $\frac{2}{3}$. So, the value of k can be, 18, 9, 6, 3, 2, or 1. Any of these values may be selected as the correct answer. Now, subtract 1 from 18: $18 - 1 = 17$.

17) The answer is $\$1,200$.

Use the simple interest formula: $I = prt$, (I = interest, p = principal, r = rate, t = time).

Simple interest $I = 2{,}500 \times 0.08 \times 6 = 1{,}200$.

She will pay $1,200 in interest at the end of 6 years.

18) The answer is $1 \pm \sqrt{x+1}$.

Rewrite the quadratic equation as the vertex form:

$f(x) = x^2 - 2x \rightarrow f(x) = (x^2 - 2x + 1) - 1 \rightarrow f(x) = (x-1)^2 - 1.$

Now, replace $f(x)$ with y: $y = (x-1)^2 - 1$.

Then, replace all x's with y and all y's with x:

$x = (y-1)^2 - 1.$

Solve for y:

$x = (y-1)^2 - 1 \rightarrow x + 1 = (y-1)^2 \rightarrow \sqrt{x+1} = |y-1|$

$\rightarrow y - 1 = \pm\sqrt{x+1} \rightarrow y = 1 \pm \sqrt{x+1}.$

Finally, replace y with $f^{-1}(x)$:

$f^{-1}(x) = 1 \pm \sqrt{x+1}$

19) The answer is 2 and -2.

The logarithm is another way of writing an exponent. $log_b\, y = x$ is equivalent to $y = b^x$.

Rewrite the logarithm in exponent form:

$log_3(x^2 + 5) = 2 \rightarrow x^2 + 5 = 3^2 \rightarrow x^2 + 5 = 9 \rightarrow x^2 = 4 \rightarrow x = 2$ or $x = -2$.

Verify the solutions:

For $x = 2$, $log_3((2)^2 + 5) = log_3(4+5) = log_3 9 = 2$,

For $x = -2$, $log_3((-2)^2 + 5) = log_3(4+5) = log_3 9 = 2$.

Therefore, there are two solutions to this equation.

20) The answer is 84.

The description $8 + 2x$ is 16 more than 20 can be written as the equation $8 + 2x = 16 + 20$, which is equivalent to $8 + 2x = 36$. Subtracting 8 from each side of $8 + 2x = 36$ gives $2x = 28$. Since $6x$ is 3 times $2x$, multiplying both sides of $2x = 28$ by 3 gives $6x = 84$.

21) The answer is 8.

$log_x 2 = \frac{1}{3}$ (log rule: $log_a(b) = \frac{1}{log_b(a)}$), $\frac{1}{log_2 x} = \frac{1}{3}$. Using cross multiplication:

$1 \times 3 = 1 \times log_2 x \rightarrow 3 = log_2 x.$

(Using logarithmic definition: $log_a b = c \rightarrow b = a^c$) $x = 2^3 \rightarrow x = 8.$

22) The answer is $y = 0$.

In a rational function, if the denominator has a higher degree than the numerator, the horizontal asymptote is the x −axis or the line $y = 0$. In the function $f(x) = \frac{x+2}{x^2+1}$, the degree of the numerator is 1 (x to the power of 1) and the degree of the denominator is 2 (x to the power of 2). Therefore, the horizontal asymptote is the line $y = 0$.

23) There is no translation of $f(x)$.

First, rewrite the equation in expanded form as $f(x) = (x-1)(x^2-4) = x^3 - x^2 - 4x + 4$. Since the equation is a polynomial function of degree three, and the leading coefficient is positive, then $f(x) \to -\infty$ as $x \to -\infty$ and $f(x) \to +\infty$ as $x \to +\infty$. On the other hand, polynomial functions are continuous. Necessarily, $f(x)$ has a real root on the domain, and considering that the translation of an arbitrary function does not change the end behavior of the function and continuity, there is no translation of $f(x)$.

24) The answer is $\frac{10}{3} < x < 10$.

$\left|\frac{x}{2} - 2x + 10\right| < 5 \to \left|-\frac{3}{2}x + 10\right| < 5 \to -5 < -\frac{3}{2}x + 10 < 5$

Subtract 10 from all sides of the inequality:

$-5 - 10 < -\frac{3}{2}x + 10 - 10 < 5 - 10 \to -15 < -\frac{3}{2}x < -5$.

Multiply all sides by 2: $2 \times (-15) < 2 \times \left(-\frac{3x}{2}\right) < 2 \times (-5) \to -30 < -3x < -10$.

Divide all sides by -3 (Remember that when you divide all sides of an inequality by a negative number, the inequality sign will be swapped): $\frac{-30}{-3} > \frac{-3x}{-3} > \frac{-10}{-3} \to 10 > x > \frac{10}{3} \to \frac{10}{3} < x < 10$.

25) The answer is 28.

To solve for x, isolate the radical on one side of the equation. Divide both sides by 4. Then: $4\sqrt{2x+9} = 28 \to \frac{4\sqrt{2x+9}}{4} = \frac{28}{4} \to \sqrt{2x+9} = 7$. Square both sides:

$\left(\sqrt{(2x+9)}\right)^2 = 7^2$. Then: $2x + 9 = 49 \to 2x = 40 \to x = 20$. Substitute x by 20 in the original equation and check the answer: $x = 20 \to 4\sqrt{2(20)+9} = 4\sqrt{49} = 4(7) = 28$

26) The answer is 12.

let x be the number of gallons of water the container holds when it is full.

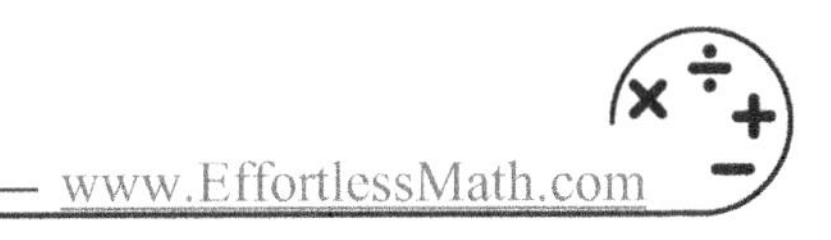

Then; $\frac{7}{24}=\frac{3.5}{x}\rightarrow x=\frac{24\times3.5}{7}=12$

27) The answer is $y=-|x|+1$.

The general form of an absolute function is: $f(x)=a|x-h|+k$

Since the graph opens downward with a slope of 1, then a is a negative one. The graph moved 1 unit up, so the value of k is 1. $y=-|x|+1$

28) The answer is $\frac{1}{8}$.

To find the coefficient of x^7, first, let $x^7=x^{2k-1}\rightarrow 2k-1=7\rightarrow 2k=8\rightarrow k=4$.

Now, put the value obtained for k in the coefficient of the general term of the series $\frac{k-1}{k!}$.

Therefore: $k=4\rightarrow\frac{4-1}{4!}=\frac{3}{24}=\frac{1}{8}$.

29) The answer is 0.

$\sqrt{x}=4\rightarrow x=16$, then; $\sqrt{x}-7=\sqrt{16}-7=4-7=-3$ and $\sqrt{x-7}=\sqrt{16-7}=\sqrt{9}=3$

Then: $(\sqrt{x-7})+(\sqrt{x}-7)=3+(-3)=0$

30) The function is positive in $\{(-6,-3)\cup(4,+\infty)\}$ and negative in $\{(-\infty,-6)\cup(-3,2)\}$.

According to the graph, the function corresponding to the graph in intervals $(-\infty,-6)$ and $(-3,2)$ is negative, because the graph of the function for every number in these intervals is under the x −axis. Similarly, the function in intervals $(-6,-3)$ and $[4,+\infty)$ is positive, as the function value is greater than zero for every number in these intervals.

31) Domain: $(-6,+\infty)-\{4\}$, and Range: $[-4,3]\cup\{5\}$.

Looking at the graph, see that the function is piecewise. So, the domain is the union of all pieces. It is clear that the domain of the graph is defined in $(-6,-3)$, $[-3,4)$ and $(4,+\infty)$. Therefore, the domain is $(-6,-3)\cup[-3,4)\cup(4,+\infty)=(-6,+\infty)-\{4\}$. Similarly, the range is the union of point 5 and the interval $[-4,3]$. That is the set of $[-4,3]\cup\{5\}$.

32) The answer is 45.

To solve for the variable, isolate it on one side of the equation. For this equation, multiply both sides by 5. Then: $\frac{3x}{5}=27\rightarrow\frac{3x}{5}\times5=27\times5\rightarrow 3x=135$

Now, divide both sides by 3. $x=\frac{135}{3}=45$

33) The answer is 2.

$81=3^{2x}$. Convert to base 3: $81=3^{2x}=3^4=3^{2x}$. If $a^{f(x)}=b^{g(x)}$, then $f(x)=g(x)$

Therefore: $4 = 2x \rightarrow x = \frac{4}{2} \rightarrow x = 2$

34) The answer is $\$20.80$.

Let x be the hourly pay in dollars and cents for the junior employee.

The salary for the senior employee who works 20 hours can be expressed as $20 \times 17 = 340$ dollars. The salary for the junior employee who works 15 hours can be expressed as $15x$ dollars. Since the senior employee earns $\$28.00$ more than the junior employee, we can set up the equation: $15x + 28.00 = 340$. Solving for x, we get $x = 20.80$.

Therefore, the hourly pay in dollars and cents for the junior employee is $\$20.80$.

35) The answer is 1.02.

The initial deposit earns 2 percent interest compounded annually. Thus, at the end of year 1, the new value of the account is the initial deposit of $\$150$ plus 2 percent of the initial deposit:

$\$150 + \frac{2}{100}(\$150) = \$150(1.02)$.

Since the interest is compounded annually, the value at the end of each succeeding year is the sum of the previous year's value plus 2 percent of the previous year's value. This is equivalent to multiplying the previous year's value by 1.02. Thus, after 2 years, the value will be $\$150(1.02)(1.02) = \$(150)(1.02)^2$; and after 3 years, the value will be $(150)(1.02)^3$; and after n years, the value will be $(150)(1.02)^n$. Therefore, in the formula for the value of Sara's account after n years $(150)(x)^n$, the value of x is 1.02.

ALEKS Mathematics Practice Test 3

1) The answer is $x^{\frac{21}{4}}$.

When multiplying two exponents: $(x^6)^{\frac{7}{8}} = x^{6 \times \frac{7}{8}} = x^{\frac{42}{8}} = x^{\frac{21}{4}}$

2) The answer is 0.

$g(x) = |2x| \rightarrow g(-2) = |2(-2)| = |-4| = 4$

$f(x) = -x + 4 \rightarrow f(4) = -4 + 4 = 0$

3) The answer is 5.

To calculate the average rate of change in weight, put $a = 2003$ and $b = 2021$ with the corresponding values $f(a) = 80$ and $f(b) = 170$. In this case, using this formula $\frac{f(b)-f(a)}{b-a}$, and substituting the values, we have: The average rate of change $= \frac{170-80}{2021-2003} = \frac{90}{18} = 5$.

4) The answer is $y = \frac{x-3}{5}$.

Let y be Anna's age: $5y + 3 = x \rightarrow 5y = x - 3 \rightarrow y = \frac{x-3}{5}$

5) The answer is -2.

Use the geometric sequence formula: $a_n = a_1 r^{n-1} \rightarrow a_n = a_{n-1} r$.

Now, substitute $a_2 = \frac{1}{2}$ and $a_3 = -\frac{1}{8}$. So, $a_3 = a_2 r = -\frac{1}{8} = \left(\frac{1}{2}\right) r \rightarrow r = -\frac{1}{4}$. Then,

$x = a_1 \rightarrow a_2 = x\left(-\frac{1}{4}\right) = \frac{1}{2} \rightarrow x = -2$.

6) The answer is $3i + 2$.

Use the FOIL (First-Out-In-Last) method:

$i(4 - 5i) + i - (2i + 3) = 4i - 5(i)^2 + i - 2i - 3$. Combine like terms:

$4i - 5(i)^2 + i - 2i - 3 = 3i - 3 - 5i^2$.

Since $i = \sqrt{-1} \rightarrow i^2 = -1$, then: $3i - 3 - 5i^2 = 3i - 3 - 5(-1) = 3i + 2$.

7) The answer is -12.

Isolate x in the equation and solve. Then:

$\frac{3}{4}(x - 2) = 3\left(\frac{1}{6}x - \frac{3}{2}\right)$, expand $\frac{3}{4}$ and 3 to the parentheses $\rightarrow \frac{3}{4}x - \frac{3}{2} = \frac{1}{2}x - \frac{9}{2}$. Add $\frac{3}{2}$ to both sides: $\frac{3}{4}x - \frac{3}{2} + \frac{3}{2} = \frac{1}{2}x - \frac{9}{2} + \frac{3}{2}$.

Simplify: $\frac{3}{4}x = \frac{1}{2}x - 3$. Now, subtract $\frac{1}{2}x$ from both sides:

$\frac{3}{4}x - \frac{1}{2}x = \frac{1}{2}x - 3 - \frac{1}{2}x$. Simplify: $\frac{1}{4}x = -3$. Multiply both sides by 4: $(4)\frac{1}{4}x = -3(4)$, simplify

$x = -12$

8) The answer is 161.

To find the solution to this problem, first, find the number of common elements between H and N. It means that $23 + 3 = 26$. Considering that the set P has 3 elements in common with the other two sets, we have to subtract 3 from the set of all members of P because the repetition of the member should not occur. Therefore, $84 + 10 + 41 + 26 = 161$.

9) The answer is 7.

Adding 6 to each side of the inequality $4n - 3 \geq 1$ yields the inequality $4n + 3 \geq 7$. Therefore, the least possible value of $4n + 3$ is 7.

10) The answer is 6.

The four-term polynomial expression can be factored completely, by grouping, it as follows:

$(x^3 - 6x^2) + (3x - 18) = 0,\ x^2(x - 6) + 3(x - 6) = 0,\ (x - 6)(x^2 + 3) = 0.$

By the zero product property, set each factor of the polynomial equal to 0 and solve each resulting equation for x. This gives $x = 6$ or $x = \pm i\sqrt{3}$, respectively. Because the equation the question asks for the real value of x that satisfies the equation, the correct answer is 6.

11) The answer is $-\frac{1}{2}$.

To find the slope of a line, find two points on the line and use the slope formula: $m = \frac{y_2 - y_1}{x_2 - x_1}$

Let's choose these two points: $(-4, 0)$ and $(0, -2)$. The slope of the line is:

$$m = \frac{y_2 - y_1}{x_2 - x_1} = \frac{-2 - 0}{0 - (-4)} = \frac{-2}{4} = -\frac{1}{2}$$

The slope of the line is $-\frac{1}{2}$.

12) The answer is $45{,}000$.

Three times 18,000 is 54,000. One-sixth of them canceled their tickets. One-sixth of 54,000 equals 9,000 ($\frac{1}{6} \times 54{,}000 = 9{,}000$).

45,000 ($54{,}000 - 9{,}000 = 45{,}000$) fans are attending this week.

13) The answer is $\frac{n(n-1)(n+7)}{3}$.

First, by using the property $\sum_{i=1}^{n}(a_i + b_i) = \sum_{i=1}^{n} a_i + \sum_{i=1}^{n} b_i$, we have:

$\sum_{i=1}^{n}(i^2 + 3i - 4) = \sum_{i=1}^{n} i^2 + \sum_{i=1}^{n} 3i + \sum_{i=1}^{n} -4$.

Now, use the formula: $\sum_{i=1}^{n} ca_i = c\sum_{i=1}^{n} a_i \rightarrow \sum_{i=1}^{n} 3i = 3\sum_{i=1}^{n} i$. So,

$\sum_{i=1}^{n}(i^2 + 3i - 4) = \sum_{i=1}^{n} i^2 + 3\sum_{i=1}^{n} i + \sum_{i=1}^{n} -4$.

Since, $\sum_{i=1}^{n} i^2 = \frac{n(n+1)(2n+1)}{6}$, and $\sum_{i=1}^{n} i = \frac{n(n+1)}{2}$, and $\sum_{i=1}^{n} c = nc$, Therefore:

$\sum_{i=1}^{n}(i^2 + 3i - 4) = \frac{n(n+1)(2n+1)}{6} + 3\frac{n(n+1)}{2} - 4n = \frac{(2n^3+3n^2+n)+9(n^2+n)-24n}{6}$.

Once simplified the answer is:

$\sum_{i=1}^{n}(i^2 + 3i - 4) = \frac{n(n-1)(n+7)}{3}$.

14) The answer is $(-1, 3)$ and $(2, 2)$.

To solve this system of equation, expand the equations and simplify as follows:

$$\begin{cases} 3y + 4x = 3x^2 + 2 \\ x + 2 = 3y - 2(x-1)^2 \end{cases} \rightarrow \begin{cases} 3y = 3x^2 - 4x + 2 \\ 3y = 2x^2 - 3x + 4 \end{cases}$$

Substitute the first equation for $3y$ in the second equation.

$3x^2 - 4x + 2 = 2x^2 - 3x + 4 \rightarrow x^2 - x - 2 = 0$,

Now, solve this quadratic equation $x^2 - x - 2 = 0$. Use the polynomial identity as $(x + a)(x + b) = x^2 + (a + b)x + ab$, then, $x^2 - x - 2 = (x + 1)(x - 2)$. Therefore,

$x^2 - x - 2 = 0 \rightarrow (x + 1)(x - 2) = 0 \rightarrow x = -1$ and $x = 2$.

Finally, to find y, put the obtained values for x in one of the equations. So,

$x = -1 \rightarrow (-1) + 2 = 3y - 2(-1 - 1)^2 \rightarrow 1 = 3y - 8 \rightarrow y = 3$,

$x = 2 \rightarrow (2) + 2 = 3y - 2(2 - 1)^2 \rightarrow 4 = 3y - 2 \rightarrow y = 2$.

The ordered pairs of $(-1,3)$ and $(2,2)$ are the solution of this system of equations.

15) The answer is $t \leq \frac{5,000}{r}$.

Let t be the time taken to climb the mountain per hour. According to the content of the question, we know that r is the average speed of climbing the mountain in feet per hour.

By multiplying the speed r by the time spent climbingthe mountain $r \times t$, the distance traveled is obtained.

Since the maximum height of the mountain is 5,000 feet, therefore the traversed height $r \times t$ must be less than or equal to 5,000. That is, $rt \leq 5{,}000$ or $t \leq \frac{5{,}000}{r}$.

16) The answer is 1.

The function $f(x)$ is undefined when the denominator of $\frac{1}{(x-3)^2+4(x-3)+4}$ is equal to zero. The expression $(x-3)^2 + 4(x-3) + 4$ is a perfect square.

$(x-3)^2 + 4(x-3) + 4 = ((x-3)+2)^2$ which can be rewritten as $(x-1)^2$. The expression $(x-1)^2$ is equal to zero if and only if $x = 1$. Therefore, the value of x for which $f(x)$ is undefined is 1.

17) The answer is 5.

Simple interest (I) is calculated by multiplying the initial deposit (p), by the interest rate (r), and time (t). $(\$360 - \$240 = \$120)$

$$120 = 240 \times 0.10 \times t \rightarrow 120 = 24t \rightarrow t = \frac{120}{24} = 5$$

So, it takes 5 years to get \$360 with an investment of \$240.

18) The answer is $f^{-1}(x) = -\log_2\left(\frac{x}{5}\right)$.

First, replace $f(x)$ with y: $y = 5 \cdot \left(\frac{1}{2}\right)^x$. Then, replace all x's with y and all y's with x:

$x = 5 \cdot \left(\frac{1}{2}\right)^y$. Now, solve for y:

$$x = 5 \cdot \left(\frac{1}{2}\right)^y \rightarrow \frac{x}{5} = \left(\frac{1}{2}\right)^y \rightarrow ln\left(\frac{x}{5}\right) = ln\left(\frac{1}{2}\right)^y \rightarrow ln\left(\frac{x}{5}\right) = y\, ln\left(\frac{1}{2}\right)$$

$$\rightarrow ln\left(\frac{x}{5}\right) = -y\, ln\, 2 \rightarrow y = -\frac{ln\left(\frac{x}{5}\right)}{ln\, 2}.$$

Finally, replace y with:

$$f^{-1}(x) = -\frac{ln\left(\frac{x}{5}\right)}{ln\, 2} \text{ or } f^{-1}(x) = -\log_2\left(\frac{x}{5}\right).$$

19) The answer is 3.

Since we are dealing with an absolute value, $f(a) = 20$ means that either $11 + a^2 = 20$ or $11 + a^2 = -20$. Let's start with the positive value (20) and see what we get. If $11 + a^2 = 20$, then $a^2 = 9$. Taking the square root, we get $a = 3\ or - 3$, On the other hand, if $11 + a^2 = -20$, then: $a = \sqrt{-31}$, notice that the question states that a is a positive integer, therefore the answer is 3.

20) The answer is $A = \frac{100x}{y}$.

Let the number be A. Then: $x = y\% \times A$. Solve for A. $x = \frac{y}{100} \times A$

Multiply both sides by $\frac{100}{y}$: $x \times \frac{100}{y} = \frac{y}{100} \times \frac{100}{y} \times A \rightarrow A = \frac{100x}{y}$

21) The answer is 360.

One of the four numbers is x; let the other three numbers be y, z, and w. Since the sum of four numbers is 600, the equation $x + y + z + w = 600$ is true. The statement that x is 50% more than the sum of the other three numbers can be represented as

$x = 1.5(y + z + w)$ or $\frac{x}{1.5} = y + z + w \rightarrow \frac{2x}{3} = y + z + w$

Substituting the value $y + z + w$ in the equation $x + y + z + w = 600$

gives $x + \frac{2x}{3} = 600 \rightarrow \frac{5x}{3} = 600 \rightarrow 5x = 1{,}800 \rightarrow x = \frac{1{,}800}{5} = 360$.

22) The answer is $b = 570$.

This is a simple matter of substituting values for variables.

We are given that the 50 cars were washed today, therefore we can substitute that for a.

Giving us the expression $\frac{40(50)-500}{50} + b$. We are also given that the profit was \$600, which we can substitute for $f(a)$. Which gives us the equation $600 = \frac{40(50)-500}{50} + b$

Simplifying the fraction gives us the equation $600 = 30 + b$

And subtracting both sides of the equation by 30 gives us $b = 570$, which is the answer.

23) The answer is $-\frac{1}{2}$.

Considering the binomial theorem $\left(2x - \frac{c}{x}\right)^5 = \sum_{k=0}^{5} \binom{5}{k} (2x)^{5-k} \left(-\frac{c}{x}\right)^k$. Rewrite, as $\left(2x - \frac{c}{x}\right)^5 = \sum_{k=0}^{5} \binom{5}{k} (2)^{5-k} (-c)^k x^{5-2k}$. Now, multiply x on both sides of the above equation.

$x\left(2x - \frac{c}{x}\right)^5 = x\left(\sum_{k=0}^{5} \binom{5}{k} (2)^{5-k} (-c)^k x^{5-2k}\right) = \sum_{k=0}^{5} \binom{5}{k} (2)^{5-k} (-c)^k x^{6-2k}$.

Then, the constant term is a value where $6 - 2k = 0 \rightarrow 2k = 6 \rightarrow k = 3$. Substitute $k = 3$ in general term of series $\binom{5}{k} (2)^{5-k} (-c)^k x^{6-2k}$, and it is equal to the constant term.

$\binom{5}{3} (2)^{5-3} (-c)^3 x^{6-2\times 3} = \frac{5!}{2!3!} 2^2 = -40c^3$.

Evaluate $-40c^3 = 5 \rightarrow c^3 = -\frac{1}{8} \rightarrow c = -\frac{1}{2}$.

24) The answer is $8\sqrt{2}$.

Find the factor of numbers: $5\sqrt{2}+\sqrt{32}-2^{\frac{1}{2}}$

$32 = 2\times2\times2\times2\times2 = 2^4\times2$

Use the radical rules: $\sqrt[n]{a^n}=a$ and $\sqrt{ab}=\sqrt{a}\times\sqrt{b}$. Then:

$\sqrt{32}=\sqrt{2\times2\times2\times2\times2}=\sqrt{2^4}\times\sqrt{2}=2^2\sqrt{2}=4\sqrt{2}$

Now, use the rule: $\sqrt[n]{x}=x^{\frac{1}{n}}$. Therefore: $2^{\frac{1}{2}}=\sqrt{2}$.

Finally: $5\sqrt{2}+\sqrt{32}-2^{\frac{1}{2}}=5\sqrt{2}+4\sqrt{2}-\sqrt{2}=8\sqrt{2}$.

25) The answer is $\frac{x}{3}$.

Plug in $\frac{z}{3}$ for z and simplify.

$$x=\frac{8y+\frac{r}{r+1}}{\frac{6}{\frac{z}{3}}}=\frac{8y+\frac{r}{r+1}}{\frac{3\times6}{z}}=\frac{8y+\frac{r}{r+1}}{3\times\frac{6}{z}}=\frac{1}{3}\times\frac{8y+\frac{r}{r+1}}{\frac{6}{z}}=\frac{x}{3}$$

26) The answer is $3c$.

If $a=b+2c$ and $b=c$, then it follows logically that $a=b+2c\rightarrow a=c+2c=3c$.

27) The answer is $\frac{3}{2}$.

Plug $x=0$ into the equation of the exponential function:

$f(0)=2^{3(0)-1}+1=2^{0-1}+1=2^{-1}+1=\frac{1}{2}+1=\frac{3}{2}$.

28) The answer is $y=5$.

Solve the system of equations by the elimination method.

$\begin{array}{l} 3x-4y=-20 \\ -x+2y=10 \\ \hline \end{array}$ $\rightarrow$ Multiply the second equation by 3, then add it to the first equation.

$\begin{array}{l} 3x-4y=-20 \\ 3(-x+2y=10) \\ \hline \end{array} \Rightarrow \begin{array}{l} 3x-4y=-20 \\ -3x+6y=30 \end{array} \Rightarrow$ Add the equations $2y=10\Rightarrow y=5$

29) The answer is x.

$$f(x)=\frac{1}{2}x+2\rightarrow f(2x-4)=\frac{1}{2}(2x-4)+2=\left(\frac{1}{2}\right)(2x)+\left(\frac{1}{2}\right)(-4)+2=x-2+2=x$$

30) The answer is $b-3$.

Isolate c in the equation: $\frac{1}{b-1}=\frac{1}{c+2}\rightarrow c+2=b-1\rightarrow c=b-1-2\rightarrow c=b-3$.

31) The answer is $100,000$.

Solve for x: $0.00104 = \frac{104}{x}$, multiply both sides by x, $(0.00104)(x) = \frac{104}{x}(x)$.

Simplify $0.00104x = 104$. Divide both sides by 0.00104: $\frac{0.00104x}{0.00104} = \frac{104}{0.00104}$, simplify

$x = \frac{104}{0.00104} = 100{,}000$

32) The answer is 6.88.

$$\frac{1\frac{4}{3}+\frac{1}{4}}{2\frac{1}{2}-\frac{17}{8}} = \frac{\frac{7}{3}+\frac{1}{4}}{\frac{5}{2}-\frac{17}{8}} = \frac{\frac{28+3}{12}}{\frac{20-17}{8}} = \frac{\frac{31}{12}}{\frac{3}{8}} = \frac{31\times 8}{12\times 3} = \frac{31\times 2}{3\times 3} = \frac{62}{9} \approx 6.88$$

33) The answer is $-\frac{5}{3}$.

To find the vertical asymptote(s) of a rational function, set the denominator equal to 0 and solve for x. Then: $3x + 5 = 0 \rightarrow 3x = -5 \rightarrow x = -\frac{5}{3}$.

The vertical asymptote is $x = -\frac{5}{3}$.

34) The answer is $\frac{m(m+2)}{2}$.

In Pascal's triangle, we know that the last row had $m + 1$ entries. The row above it had m entries, and each row above had 1 fewer entries than the row just below it. There is 1 triangle in the top row. The arithmetic sequence is obtained starting from 1 to $m + 1$. Using the arithmetic sequence: $S_n = \frac{1}{2}n(a_1 + a_n)$. Therefore,

$S_m = \frac{1}{2}m(a_1 + a_m) = \frac{1}{2}m(1 + m + 1) = \frac{1}{2}m(m + 2) \rightarrow S_m = \frac{m(m+2)}{2}$.

35) The answer is $-3 \leq x \leq 5$.

The possible x values are between -3 and 5. Domain: $-3 \leq x \leq 5$

ALEKS Mathematics Practice Test 4

1) The answer is $28x + 6$.

If $f(x) = 3x + 4(x + 1) + 2$, then find $f(4x)$ by substituting $4x$ for every x in the function. This gives: $f(4x) = 3(4x) + 4(4x + 1) + 2$.

It simplifies to: $f(4x) = 3(4x) + 4(4x + 1) + 2 = 12x + 16x + 4 + 2 = 28x + 6$.

2) The answer is $-57 + 7i$.

We know that: $i = \sqrt{-1} \Rightarrow i^2 = -1$,

$(-4 + 9i)(3 + 5i) = -12 - 20i + 27i + 45i^2 = -12 + 7i - 45 = -57 + 7i$.

3) The answer is $\frac{-2x^6}{y}$.

Using rules of exponents, start in the numerator with $(-2x^2y^2)^3$, which is $(-2)^3(x^2)^3(y^2)^3$, which simplifies to $-8x^6y^6$. That is multiplied by $3x^3y$, giving $-24x^9y^7$.

4) Next, divide $\frac{-24x^9y^7}{12x^3y^8}$ to get $\frac{-2x^6}{y}$. **The answer is $\{x|x \leq -\frac{14}{3}\}$.**

Start by using the distributive property to simplify the left side of the inequality and combine like terms to get $-2x - 8 \geq x + 6$. To isolate x, subtract x and add 8 to both sides. This gives $-3x \geq 14$. To isolate the x, divide both sides by -3. Dividing by the negative changes the relationship between the sides and gives $x \leq -\frac{14}{3}$. In set-builder notation, the solution is: $\{x|x \leq -\frac{14}{3}\}$

5) The answer is $-\frac{3}{2}$.

Simplify the equation $Q(x) = 2x^3 - x^2 - 6x$. First, factor the equation: $2x^3 - x^2 - 6x = x(2x^2 - x - 6)$. To find the zeros, each factor should be equal to zero: $x(2x^2 - x - 6) = 0$. Therefore, the zeros are $x = 0$, $2x^2 - x - 6 = 0$.

At this point, evaluate the discriminant of the quadratic equation $2x^2 - x - 6 = 0$. The expression $\Delta = b^2 - 4ac$ is called the discriminant for the standard form of the quadratic equation of the form $ax^2 + bx + c = 0$. So, $\Delta = b^2 - 4ac \rightarrow \Delta = (-1)^2 - 4(2)(-6) = 49$.

Since $\Delta > 0$, the quadratic equation has two distinct solutions. Now, we will use the quadratic formula: $x_{1,2} = \frac{-b \pm \sqrt{\Delta}}{2a}$

Then, the roots are $x_1 = \frac{-(-1)+\sqrt{49}}{2(2)} = \frac{1+7}{4} = 2$, and $x_2 = \frac{-(-1)-\sqrt{49}}{2(2)} = \frac{1-7}{4} = -\frac{3}{2}$.

6) The answer is 4.

Since there is a linear relationship between the data. So, the rate of change in this model is the same value between all points. By using the rate of change formula $\frac{f(b)-f(a)}{b-a}$, for every a and b such that $a < b$. We can evaluate the rate of change for points 1 and 3.

Therefore, $\frac{21-13}{3-1} = \frac{8}{2} = 4$.

7) The answer is -21.

To find the answer to the problem, solve the equation for $r(x) = 0: \frac{4}{7}x + 12 = 0$.

Subtract 12 from both sides: $\frac{4}{7}x = -12$. Multiply both sides by $\frac{7}{4}$: $x = -21$.

8) The answer is 4.

The sum of the given expressions is $(-4x^2 + 3x - 24) + (7x^2 - 8x + 18)$. Combining like terms yields $3x^2 - 5x - 6$. Based on the form of the given equation, $a = 3$, $b = -5$, and $c = -6$. Therefore, $a + b - c = 3 + (-5) - (-6) = 4$.

9) The answer is 3 and 4.

First, factor the function: $(x - 4)(x - 3)$. To find the zeros, $f(x)$ should be zero:

$f(x) = (x - 4)(x - 3) = 0$.

Therefore, the zeros are, $(x - 4) = 0 \Rightarrow x = 4$, $(x - 3) = 0 \Rightarrow x = 3$.

10) The answer is $(x - 2)(x^2 + 2x + 4)$.

The formula for factoring the difference between two cubes is:

$(a^3 - b^3) = (a - b)(a^2 + ab + b^2)$. Here, we have $x^3 - 8 = x^3 - 2^3$. Replacing a with x and b with 2 gives: $(x - 2)(x^2 + 2x + 4)$

11) The answer is 42.

In order to find the y −intercept, you must put the value of the variable x in the equation of the function $k(x)$ equal to zero. Therefore, when $x = 0$, the y −intercept is obtained.

$x = 0 \rightarrow k(0) = 42\left(\frac{4}{5}\right)^0 = 42$.

12) The answer is 12.

Change 9^n into a base 3 exponential. Since $9 = 3^2$, you can substitute 3^2 for 9.

$3^m.(3^2)^n = 3^{12} \rightarrow 3^m.3^{2n} = 3^{12} \rightarrow 3^{m+2n} = 3^{12} \rightarrow m + 2n = 12$.

13) The answer is $16 - 4\pi$.

Subtract the area of the circle from the area of the square and divide the result by 4. The side of the square is equal to the diameter of the circle, so each side is equal to 8.

Area of the square → $8 \times 8 = 64$. Area of the circle → $A = \pi r^2 = (4 \times 4)\pi = 16\pi$

Area of the shaded region → $\frac{64 - 16\pi}{4} = \frac{64}{4} - \frac{16\pi}{4} = 16 - 4\pi$

14) The answer is 20.

Put $x = 2$ in the equation:

$q(2) = 3(2-5)^2 - 7 = 3(-3)^2 - 7 = 3(9) - 7 = 20.$

15) The answer is $35x^2 + 24xy + 4y^2$.

Use the FOIL (First, Out, In, Last) method:

$(7x + 2y)(5x + 2y) = 35x^2 + 14xy + 10xy + 4y^2$

$= 35x^2 + 24xy + 4y^2$

16) The answer is 7.54×10^1.

$(2.9 \times 10^6) \times (2.6 \times 10^{-5}) = (2.9 \times 2.6) \times (10^6 \times 10^{-5}) = 7.54 \times \left(10^{6+(-5)}\right)$

$= 7.54 \times 10^1$

17) The answer is $-2 - \frac{7c}{9b}$.

To write this expression as a single fraction, we need to find a common denominator.

The common denominator of $9b$ and $6b$ is $18b$. Then:

$-2 + \frac{3b-4c}{9b} - \frac{2b+2c}{6b} = \frac{-2(18b)}{18b} + \frac{2(3b-4c)}{18b} - \frac{3(2b+2c)}{18b}.$

Now, simplify the numerators and combine them:

$\frac{-2(18b)}{18b} + \frac{2(3b-4c)}{18b} - \frac{3(2b+2c)}{18b} = \frac{-36b}{18b} + \frac{6b-8c}{18b} - \frac{6b+6c}{18b} = \frac{-36b+6b-8c-6b-6c}{18b} = \frac{-36b-14c}{18b}.$

Divide both the numerator and denominator by 2. Then: $\frac{-36b-14c}{18b} = \frac{-18b-7c}{9b} = -2 - \frac{7c}{9b}.$

18) The answer is 1.

Rewrite the equation $x^2 - 3x + 1 = x - 3$ by simplifying. First, subtract x from both sides: $x^2 - 3x + 1 - x = x - 3 - x \rightarrow x^2 - 4x + 1 = -3$. Then, add 3 to both sides of the equation: $x^2 - 4x + 1 + 3 = -3 + 3 \rightarrow x^2 - 4x + 4 = 0$.

Now, remember that there can be 0, 1, or 2 solutions to a quadratic equation. In standard form, a quadratic equation is written as $ax^2 + bx + c = 0$.

For the quadratic equation, the expression $\Delta = b^2 - 4ac$ is called the discriminant. If the discriminant is positive, there are 2 distinct solutions for the quadratic equation. If the discriminant is 0, there is one solution for the quadratic equation and if it is negative the equation does not have any solutions.

Find the value of the discriminant: $\Delta = b^2 - 4ac \rightarrow \Delta = (-4)^2 - 4(1)(4) = 16 - 16 = 0$.

Since the discriminant is zero, the quadratic equation has one distinct solution.

19) The answer is 3.91×10^2.

To multiply two numbers in scientific notation, multiply their coefficients and add their exponents. For these two numbers in scientific notation, multiply the coefficients: $1.7 \times 2.3 = 3.91$. Add the powers of 10: $10^9 \times 10^{-7} = 10^{9+(-7)} = 10^2$. Then:

$(1.7 \times 10^9) \times (2.3 \times 10^{-7}) = (1.7 \times 2.3) \times (10^9 \times 10^{-7}) = 3.91 \times \left(10^{9+(-7)}\right)$

$= 3.91 \times 10^2$

20) The answer is $s = \frac{h-k}{t} + 25t$.

Starting with the original equation, $h = -25t^2 + st + k$, in order to get s in terms of the other variables, $-25t^2$ and k need to be subtracted from each side. This yields $st = h + 25t^2 - k$, which when divided by t will give s in terms of the other variables. The equation $s = \frac{h+25t^2-k}{t}$, can be further simplified. Another way to write the previous equations is $s = \frac{h-k}{t} + \frac{25t^2}{t}$, which can be simplified to $s = \frac{h-k}{t} + 25t$.

21) The answer is $10x + 6xy + 4x^2 + 8xz$.

Use the distributive property: $2x(5 + 3y + 2x + 4z) = 10x + 6xy + 4x^2 + 8xz$.

22) The answer is $\frac{10}{3}$.

We know that: $log_a b - log_a c = log_a \frac{b}{c}$ and $log_a b = c \rightarrow b = a^c$. Then:

$log_4(x + 2) - log_4(x - 2) = 1 \rightarrow log_4 \frac{x+2}{x-2} = 1 \rightarrow \frac{x+2}{x-2} = 4^1 = 4 \rightarrow x + 2 = 4(x - 2)$.

Therefore: $x + 2 = 4x - 8 \rightarrow 4x - x = 8 + 2 \rightarrow 3x = 10 \rightarrow x = \frac{10}{3}$.

23) The answer is 4.

To find the number of tickets the customer purchased, we can use the given function and solve for t: $c = 24.50t + 5.25$ (Substitute $c = \$103.25$) $\rightarrow 103.25 = 24.50t + 5.25$

(Subtract 5.25 from both sides) $98 = 24.5t$

(Divide both sides by 24.5) $t = \frac{98}{24.5} = 4$. Therefore, the customer purchased 4 tickets.

24) The answer is 26.

Let c be the number of students in Mr. Anderson's class. The conditions described in the question can be represented by the equations $n = 3c + 5$ and $n + 21 = 4c$. Substituting $3c + 5$ for n in the second equation gives $3c + 5 + 21 = 4c$, which can be solved to find $c = 26$.

25) The answer is -4.

To find the x −intercept, put $x = 0$ in the equation $3x - 6y = 24$, then we get:

$x - 0 \rightarrow 3(0) - 6y = 24 \rightarrow -6y = 24 \rightarrow y = -4$.

So, the y −intercept is -4.

26) The answer is 22.

If there is a small semicircle inside a big semicircle, then: the perimeter of the big semicircle = $\frac{8\times\pi}{2} = \frac{24}{2} = 12\ cm$. The perimeter of the small semicircle $= \frac{4\times\pi}{2} = \frac{12}{2} = 6\ cm$.

Total perimeter $= 4 + 6 + 12 = 22\ cm$.

27) The answer is $f^{-1}(x) = \pm\sqrt{x-1}$.

First, replace $f(x)$ with y: $y = x^2 + 1$. Then, replace all x's with y and all y's with x: $x = y^2 + 1$. Now, solve for y: $x = y^2 + 1 \rightarrow x - 1 = y^2 \rightarrow |y| = \sqrt{x-1} \rightarrow y = \pm\sqrt{x-1}$. Finally, replace y with $f^{-1}(x)$: $f^{-1}(x) = \pm\sqrt{x-1}$.

28) The answer is $3\sqrt{2}$.

Find the factor of the numbers:

$8 = 4 \times 2 = 2^2 \times 2$

$50 = 25 \times 2 = 5^2 \times 2$

$72 = 36 \times 2 = 6^2 \times 2$

Now use the radical rule: $\sqrt[n]{a^n} = a$.

Finally: $\sqrt{8} - \sqrt{50} + \sqrt{72} = \sqrt{2^2 \times 2} - \sqrt{5^2 \times 2} + \sqrt{6^2 \times 2} = 2\sqrt{2} - 5\sqrt{2} + 6\sqrt{2} = 3\sqrt{2}$.

29) The answer is 3 or 6.

The x –intercepts of the parabola represented by $y = x^2 - 9x + 18$ in the xy –plane are the values of x for which y is equal to 0. The factored form of the equation, $y = (x - 3)(x - 6)$, shows that y equals 0 if and only if $x = 3$ or $x = 6$. Thus, the x –intercepts of the parabola are 3 and 6.

30) There is no solution.

To solve this system of equations, we can use the method of elimination. If we multiply the first equation by -3, we get $-12x + 21y = 6$. If we add this equation to the second equation, we get $12x - 21y + (-12x + 21y) = -42 + (6) \rightarrow 0 = -36$.

This equation simplifies to $0 = -36$, which is not true. Therefore, there is no solution to this system of equations.

31) The answer is $(-\infty, +\infty)$.

Since the equation of the function f is quadratic, the domain of this function is all real numbers. It means that the interval $(-\infty, +\infty)$ and is represented by $\mathbb{R}$.

32) The answer is 60.

To calculate the average rate of change in the exponential growth $f(x)$, considering that the given interval $1 \leq x \leq 3$, put $a = 1$ and $b = 3$ with the corresponding values $f(a) = 15$ and $f(b) = 135$. Now, use this formula $\frac{f(b)-f(a)}{b-a}$, and substituting the values:

The average rate of change $= \frac{135-15}{3-1} = \frac{120}{2} = 60$.

33) The answer is $\frac{n(n+5)}{2}$.

Use the formula $\sum_{i=1}^{n}(a_i + b_i) = \sum_{i=1}^{n} a_i + \sum_{i=1}^{n} b_i$, so:

$\sum_{i=1}^{n}(i + 2) = \sum_{i=1}^{n} i + \sum_{i=1}^{n} 2$.

Now, using these properties $\sum_{i=1}^{n} i = \frac{n(n+1)}{2}$, and $\sum_{i=1}^{n} c = nc$. Therefore,

$$\sum_{i=1}^{n}(i + 2) = \frac{n(n+1)}{2} + 2n = \frac{n^2+n+4n}{2} = \frac{n^2+5n}{2} = \frac{n(n+5)}{2}.$$

34) The answer is three.

According to the given graph, the value of the function is zero for two inputs. It means that the degree of the function is greater than 2. On the other hand, on the left $f(x)$ goes to $-\infty$, and on the right $f(x)$ goes to $+\infty$, so the function of odd degree. Therefore, the degree of the function $f(x)$ is three.

35) The answer is 4.

Since the function is continuous, the function has at least one root every time it changes sign. According to the table, in the intervals of $(-1,0)$, $(0,2)$, $(2,5)$, and $(10,18)$, the values of the function change from positive to negative and vice versa. Therefore, the function $f(x)$ has at least four zero values. So, the minimum degree is 4.

ALEKS Mathematics Practice Test 5

1) The answer is $24d + 72$.

Set a proportion: $\frac{1}{8} = \frac{3d+9}{x} \rightarrow x = 8(3d + 9) = 24d + 72$.

2) The answer is $\$30$.

If the regular price for a concert ticket is $\$100$, and a different vendor offers a 10% discount on the regular price, then the discounted price would be:

Discounted price $=$ Regular price $-$ (Discount rate $\times$ Regular price)

Discounted price $= \$100 - (0.1 \times \$100) \rightarrow$ Discounted price $= \$90$

Therefore, each ticket costs $90 from the discounted vendor.

To calculate the savings, we need to find the difference between the total cost of purchasing 3 tickets at the regular price and the total cost of purchasing 3 tickets at the discounted price:

Regular price for 3 tickets $= \$100 \times 3 = \$300$

Discounted price for 3 tickets $= \$90 \times 3 = \$270$

Savings $=$ Regular price for 3 tickets $-$ Discounted price for 3 tickets

Savings $= \$300 - \$270$ Savings $= \$30$

So, the savings in dollars and cents by purchasing 3 tickets from the discounted vendor instead of the regular vendor would be $30.

3) The answer is 7×10^2.

$\frac{1.4\times10^{-7}}{2\times10^{-10}} = \frac{1.4}{2} \times 10^{10-7} = 0.7 \times 10^3 = (7 \times 10^{-1}) \times 10^3 = 7 \times 10^{3-1} = 7 \times 10^2$

4) The answer is $y = 0.25x$.

To find the equation of the relationship of direct variation, it is enough to find the value of k. Substitute the given values of x and y into the formula $y = kx$ and evaluate the value of k. So, we get:

$\begin{matrix} x = 24 \\ y = 6 \end{matrix} \rightarrow 6 = 24k \rightarrow k = \frac{6}{24} = k = \frac{1}{4}$ or 0.25.

Next, put the obtained value in the formula: $y = 0.25x$.

5) The answer is -15.

We know $log_a \frac{x}{y} = log_a x - log_a y$. Therefore:

$x = 3\left(log_2 \frac{1}{32}\right) \rightarrow x = 3(log_2 1 - log_2 32)$. In addition, we know $log_a 1 = 0$.

Therefore: $x = -3\, log_2 32$. Considering: $log_a x^b = b \times log_a x$, and $log_a a = 1$.

Then: $log_2 32 = log_2 2^5 = 5\, log_2 2 = 5$. Finally, we have:

$x = -3\, log_2 32 \rightarrow x = -3 \times 5 \rightarrow x = -15$.

6) The answer is 13.

Based on the table provided: $g(-2) = g(x = -2) = 3 \rightarrow g(3) = g(x = 3) = -2$

$3g(-2) - 2g(3) = 3(3) - 2(-2) = 9 + 4 = 13$.

7) The answer is 4.

Plug in the values of x and y of the point $(2, 12)$ in the equation of the quadratic function. Then:

$12 = a(2)^2 + 5(2) + 10 \rightarrow 12 = 4a + 10 + 10 \rightarrow 12 = 4a + 20$

$\rightarrow 4a = 12 - 20 = -8 \rightarrow a = \frac{-8}{4} = -2 \rightarrow a^2 = (-2)^2 = 4$

8) The answer is -2.

Notice that for the function $g(x)$, if $k > 0$ then $f(x) = g(x) + k$ is shifted up by $|k|$ units, and if $k < 0$ the function $f(x) = g(x) + k$ is shifted down by $|k|$ units. The graph of the function $g(x) = x^2$ is as follows:

If $g(x)$ is shifted down by 2 units, the graph of $f(x)$ is obtained. Therefore, the value of k is -2 and the corresponding equation is $f(x) = g(x) - 2 \rightarrow f(x) = x^2 - 2$.

9) The answer is 5.

$3x + x + x - 2 = x + x + x + 8$, combining like terms on each side of the given equation yields $5x - 2 = 3x + 8$. Adding 2 to both sides of the equation $5x - 2 + 2 = 3x + 8 + 2 \rightarrow 5x = 3x + 10$. Subtracting $3x$ from both sides gives $5x - 3x = 3x + 10 - 3x \rightarrow 2x = 10$. Divide both sides of $2x = 10$ by 2 to yield $x = 5$.

10) The answer is $\frac{\sqrt{3}\, d^2}{4}$.

You can also find the height of the triangle using the relationship in $30° - 60° - 90°$ triangles. The relationship among all sides of the special right triangle $30° - 60° - 90°$ is provided in this triangle:

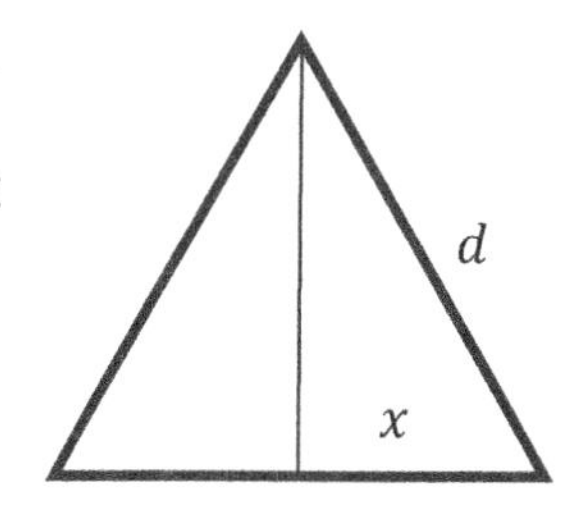

In the equilateral triangle, the side is d. Then, $2x = d \rightarrow x = \frac{d}{2}$

The height of the triangle is $\frac{d}{2} \times \sqrt{3} = \frac{d\sqrt{3}}{2}$. Then, the area of the triangle is: $\frac{1}{2}(d)\left(\frac{d\sqrt{3}}{2}\right) = \frac{\sqrt{3}\, d^2}{4}$.

The area of the equilateral triangle is $\frac{\sqrt{3}\,d^2}{4}$.

11) The answer is $4 \leq x \leq 14$.

Since this inequality contains an absolute value, then, the value inside the absolute value bars is greater than -5 and less than 5. Then:

$|x-9| \leq 5 \rightarrow -5 \leq x-9 \leq 5 \rightarrow -5+9 \leq x-9+9 \leq 5+9 \rightarrow 4 \leq x \leq 14.$

12) The answer is 9.5×10^{-4}.

The number 0.00095 is equal to 9.5×10^{-4} in scientific notation.

13) The answer is 47.

First, using $(x+y)^2 = x^2 + 2xy + y^2$. Substitute x by x^2, and y by $\frac{1}{x^2}$. We have:

$\left(x^2+\frac{1}{x^2}\right)^2 = x^4 + \frac{1}{x^4} + 2x^2 \times \frac{1}{x^2} \rightarrow \left(x^2+\frac{1}{x^2}\right)^2 = x^4 + \frac{1}{x^4} + 2.$

Substitute $x^2 + \frac{1}{x^2} = 7$. Therefore:

$(7)^2 = x^4 + \frac{1}{x^4} + 2 \rightarrow 49 = x^4 + \frac{1}{x^4} + 2 \rightarrow x^4 + \frac{1}{x^4} = 47.$

14) The answer is $2(x+1)(3x-5)$.

$6x^2 - 4x - 10 = 2(x+1)(3x-5)$

15) The answer is -9.

For the expression $(xy^{-2})^3$, use the exponential rule as $(xy)^a = x^a \times y^a$. Then:

$(xy^{-2})^3 = x^3 \times (y^{-2})^3$

In this case, by using the rule $(x^a)^b = x^{a\times b}$, we have: $(y^{-2})^3 = y^{-2\times 3} = y^{-6}$. Substitute the expression in the resulting expression from $(xy^{-2})^3$:

$(xy^{-2})^3 = x^3y^{-6}$

Considering the rule: $\left(\frac{a}{b}\right)^c = \frac{a^c}{b^c}$ for the expression $\left(\frac{y}{x}\right)^9$, then: $\left(\frac{y}{x}\right)^9 = \frac{y^9}{x^9} = y^9x^{-9}$. Substitute y^9x^{-9} and x^3y^{-6} in the composite expression: $(xy^{-2})^3\left(\frac{y}{x}\right)^9 = x^3y^{-6}y^9x^{-9}$. Arrange the terms to have the same base in the expression $x^3y^{-6}y^9x^{-9}$ to form $x^3x^{-9}y^{-6}y^9$. According to the exponential rule: $x^a \times x^b = x^{a+b}$. Thus, $x^3x^{-9}y^{-6}y^9 = x^{3-9}y^{-6+9} = x^{-6}y^3$.

Put $x^ny^m = x^{-6}y^3$. The values n and m are -6 and 3, respectively. Therefore,

$n - m = -6 - (3) = -9.$

16) The answer is $y > 0.8x - 7$.

To find the inequality equivalent to $4x - 3y < 2y + 35$, we can start by isolating y on one side of the inequality:

$4x - 3y - 2y < 2y + 35 - 2y \rightarrow 4x - 5y < 35 \rightarrow 4x - 5y - 4x < 35 - 4x$

$\rightarrow -5y < 35 - 4x$

Divide both sides of the inequality by -5. So, the equivalent inequality is $y > \frac{4}{5}x - \frac{35}{5}$, which means that the answer is $y > 0.8x - 7$.

17) The answer is 1 or 5.

The function $f(x)$ is undefined when the denominator of $\frac{1}{(x-3)^2-4}$ is equal to zero. Find the values of x for which the equation $(x-3)^2 - 4 = 0$ holds. We get:

$(x-3)^2 - 4 = 0 \rightarrow (x-3)^2 = 4 \rightarrow |x - 3| = 2$

So, $x - 3 = 2 \rightarrow x = 5$, or $x - 3 = -2 \rightarrow x = 1$.

Therefore, the value of x for which $f(x)$ is undefined is 1 or 5.

18) The answer is -8.

The problem asks for the sum of the roots of the quadratic equation $2n^2 + 16n + 24 = 0$. Dividing each side of the equation by 2 gives $n^2 + 8n + 12 = 0$. If the roots of $n^2 + 8n + 12 = 0$ are n_1 and n_2, then the equation can be factored as $n^2 + 8n + 12 = (n - n_1)(n - n_2) = 0$. Looking at the coefficient of n on each side of $n^2 + 8n + 12 = (n+6)(n+2)$ gives $n = -6$ or $n = -2$, then, $-6 + (-2) = -8$.

19) The answer is 15.

To solve the problem, plug the given information into the equation and solve for the variable b:

$$10 = \frac{(50 - 3f)}{0.5}$$

Multiplying both sides by 0.5: $5 = 50 - 3f$. Subtracting 50 from both sides: $-45 = -3f$. Dividing both sides by -3: $f = 15$. So, the construction worker completes 15 wooden fences that week.

20) The answer is $-8x^2 - 6x + 9$.

$3(-x^2 - 2x + 2) - (5x^2 - 3) = -3x^2 - 6x + 6 - 5x^2 + 3 = -8x^2 - 6x + 9$

21) The answer is $9x^2 - 30x + 25$.

Use the FOIL (First-Out-In-Last) method to simplify the expression:

$(3x-5)^2 = (3x-5)(3x-5) = 9x^2 - 15x - 15x + 25 = 9x^2 - 30x + 25.$

22) The answer is $18x^2 - 24x + 1$.

Use the polynomial identity: $(x-y)^2 = x^2 - 2xy + y^2$ for the part of the equation of function $(2-3x)^2$. Then:

$(2-3x)^2 = (2)^2 - 2(2)(3x) + (3x)^2 = 4 - 12x + 9x^2.$

Substitute the obtained expression in the equation of function and simplify.

$2(2-3x)^2 - 7 = 2(4 - 12x + 9x^2) - 7 = 18x^2 - 24x + 1.$

Therefore: $f(x) = 18x^2 - 24x + 1.$

23) The answer is 90.

In the equilateral triangle if x is the length of one side of the triangle, then the perimeter of the triangle is $3x$. Then $3x = 45 \rightarrow x = 15$ and the radius of the circle is $x = 15$. Then, the circumference of the circle is:

$2\pi r = 2\pi(15) = 30\pi, \pi = 3 \rightarrow 30\pi = 30 \times 3 = 90.$

24) The answer is $25x^2 + 9y^2 - 30xy$.

Perfect square formula: $(a-b)^2 = a^2 - 2ab + b^2$. So, $(5x-3y)^2 =$

$25x^2 - 30xy + 9y^2 = 25x^2 + 9y^2 - 30xy$

25) The answer is $\frac{ln(18)}{3}$.

If $f(x) = g(x)$, then: $ln(f(x)) = ln(g(x)) \rightarrow ln(e^{3x}) = ln(18).$

Use the logarithm rule: $log_a x^b = b\, log_a x \rightarrow ln(e^{3x}) = 3x\, ln(e) \rightarrow (3x)\, ln(e) = ln(18)$

$ln(e) = 1$, then: $(3x)ln(e) = ln(18) \rightarrow 3x = ln(18) \rightarrow x = \frac{ln(18)}{3}$

26) The answer is $\frac{3}{4} + i$.

Recall that the imaginary numbers (containing i) cannot be in the denominator of a fraction. To simplify the expression, multiply both the numerator and denominator by i.

$\frac{4-3i}{-4i} \times \frac{i}{i} = \frac{4i-3i^2}{-4i^2}, i^2 = -1$, then: $\frac{4i-3i^2}{-4i^2} = \frac{4i-3(-1)}{-4(-1)} = \frac{4i+3}{4} = \frac{4i}{4} + \frac{3}{4} = \frac{3}{4} + i$

27) The answer is $n = 2$.

Simplify and solve for n in the equation.

$2(n-1) = 3(n+2) - 10 \rightarrow 2n - 2 = 3n + 6 - 10 \rightarrow 2n - 2 = 3n - 4$

Subtract $2n$ from both sides: $-2 = n - 4$, add 4 to both sides: $n = 2$.

28) The answer is -5.

$2x^2 - 11x + 8 = -3x + 18 \rightarrow 2x^2 - 11x + 3x + 8 - 18 = 0 \rightarrow 2x^2 - 8x - 10 = 0$

$\rightarrow 2(x^2 - 4x - 5) = 0 \rightarrow$ Divide both sides by 2. Then: $x^2 - 4x - 5 = 0$, find the factors of the quadratic equation. $\rightarrow (x-5)(x+1) = 0 \rightarrow x = 5$ or $x = -1$. $a > b$, then: $a = 5$ and $b = -1$.

$\frac{a}{b} = \frac{5}{-1} = -5$

29) The answer is 29.

Here we can substitute 8 for x in the equation. Thus, $y - 3 = 2(8 + 5)$, $y - 3 = 26$.

Adding 3 to both sides of the equation: $y = 26 + 3$, $y = 29$.

30) The answer is 6.

Since the graph crosses the y −axis at $(0, r)$, then substituting 0 for x and r for y in $r = -3(0)^2 + 12(0) + 6$ creates a true statement: $r = -3(0)^2 + 12(0) + 6$, or $r = 6$.

31) The answer is $-2 \leq y \leq 6$.

The range of the function is the possible value for y. The image of the graph on the y −axis is equivalent to the range of the graph. Look at the following graph:

The interval $-2 \leq y \leq 6$ is the range of the function.

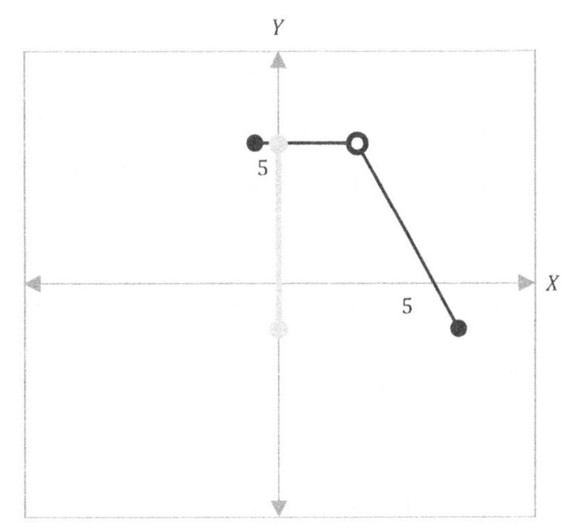

32) The answer is $(3 + 7x)(3 - 7x)$.

Rewrite the expression as $9 - 49x^2$. Use the following polynomial identity:

$x^2 - y^2 = (x + y)(x - y)$.

Then: $9 - 49x^2 = (3 + 7x)(3 - 7x)$.

33) The answer is $21x + 6$.

If $f(x) = 3x + 4(x + 1) + 2$, then find $f(3x)$ by substituting $3x$ for every x in the function. This gives: $f(3x) = 3(3x) + 4(3x + 1) + 2$

It simplifies to: $f(3x) = 3(3x) + 4(3x + 1) + 2 = 9x + 12x + 4 + 2 = 21x + 6$.

34) The answer is $\frac{3}{2}$.

To solve for the inverse function, first, replace $f(x)$ with y. Then, solve the equation for x and after that, replace every x with y and replace every y with x. Finally, replace y with $f^{-1}(x)$.

$$f(x) = \frac{10x - 3}{6} \Rightarrow y = \frac{10x - 3}{6} \Rightarrow 6y = 10x - 3 \Rightarrow 6y + 3 = 10x \Rightarrow \frac{6y + 3}{10} = x$$

$$f^{-1}(x) = \frac{6x + 3}{10} \Rightarrow f^{-1}(2) = \frac{6(2) + 3}{10} = \frac{15}{10} = \frac{3}{2}$$

35) The answer is $11\ m$ by $8\ m$.

To solve the problem, we can use the formula for the perimeter of a rectangle, which is: Perimeter $= 2 \times$ Length $+2 \times$ Width.

We are given that the perimeter of the rectangle is 38 centimeters. We are also given that the length can be represented as $(x + 6)$ and the width can be represented as $(2x - 2)$. So, we can substitute these values into the formula for the perimeter and solve for x:

$38 = 2(x + 6) + 2(2x - 2) \rightarrow 38 = 2x + 12 + 4x - 4 \rightarrow 38 = 6x + 8$

$\rightarrow 30 = 6x \rightarrow x = 5$.

Now, we can find the length and width of the rectangle by substituting $x = 5$ into the expressions for the length and width:

Length $= x + 6 = 5 + 6 = 11\ m$, and width $= 2x - 2 = 2(5) - 2 = 8\ m$.

Therefore, the dimensions of the rectangle are $11\ m$ by $8\ m$.

ALEKS Mathematics Practice Test 6

1) The answer is 160.

The equation $x + 100 = 260$ can be used to determine x, the number of dollars charged per 3 months. Subtracting 100 from both sides of this equation yields $x = 160$.

2) The answer is 20.20.

Use a calculator and calculate $\sqrt{508}$. The answer is 22.53885533 ....

Rounding the answer to the nearest hundredth, the answer is 22.54

3) The answer is 4.

If $a^y = x$, then: $log_a x = y$, therefore: $e^{ln\,4} = x \rightarrow ln\,x = ln\,4$.

Since $log_a b = log_a c \rightarrow b = c$, accordingly: $ln\,x = ln\,4 \rightarrow x = 4$.

4) The answer is $a = -3$.

To solve for a, first use the distributive property to simplify $8(a + 8)$. Then:

$8(a + 8) = 8a + 64$

Now, combine like terms: $-3a + 8(a + 8) = 49 \rightarrow -3a + 8a + 64 = 49 \rightarrow$

$5a + 64 = 49$

Subtract 64 from both sides: $5a + 64 - 64 = 49 - 64 \rightarrow 5a = -15 \rightarrow a = -3$

5) The answer is no solution.

First, find a common denominator for x and $\frac{6x}{3+x}$. It's $3 + x$. Then:

$\frac{6x}{3+x} - x = \frac{6x}{3+x} - \frac{x(3+x)}{3+x} = \frac{6x-3x-x^2}{3+x} = \frac{3x-x^2}{3+x}$.

Now, multiply both sides of equation $1 = \frac{3x-x^2}{3+x}$ by $3 + x$. Then:

$(3 + x) \times 1 = (3 + x) \times \frac{3x-x^2}{3+x}$.

Rewrite the expression: $3x - x^2 = 3 + x$. Simplify as $x^2 - 2x + 3 = 0$. Use the formula $\Delta = b^2 - 4ac$ from the standard form of the quadratic equation $ax^2 + bx + c = 0$:

$\Delta = (-2)^2 - 4(1)(3) = 4 - 12 = -8$

Because, $\Delta < 0$. Therefore, the equation has no real answer.

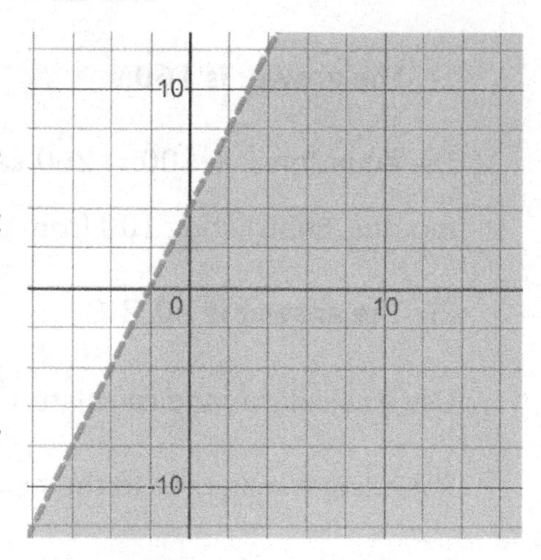

6) The answer is on the graph.

To draw the graph of inequality $-8x < 16 - 4y$, first simplify the inequality and solve for y. Solving the inequality for y: subtract 16 from both sides:

$-8x - 16 < 16 - 4y - 16 \rightarrow -8x - 16 < -4y$. Divide both sides by -4. (Remember: for dividing or multiplying both sides by negative numbers, flip the direction of the inequality sign.)

$-8x - 16 < -4y \rightarrow \frac{-8x-16}{-4} > \frac{-4y}{-4} \rightarrow 2x + 4 > y$. Now, graph the inequality $2x + 4 > y$.

To draw the graph of $y < 2x + 4$, you first need to graph the line: $y = 2x + 4$

Since there is a less than (<) sign, draw a dashed line. The slope is 2 and the y-intercept is 4. Then, choose a testing point and substitute the value of x and y from that point into the inequality. The easiest point to test is the origin: $(0, 0)$

$(0,0) \rightarrow y < 2x + 4 \rightarrow 0 < 2(0) + 4 \rightarrow 0 < 4$

This is correct! 0 is less than 4. So, this part of the line (on the right side) is the solution of this inequality.

7) The answer is $145 + 35i$.

Use the FOIL (First, Out, In, Last) method to multiply two imaginary expressions:

$(-5)(-3) + (-5)(-13i) + (10i)(-3) + (10i)(-13i) = 15 + 65i - 30i - 130i^2$

Combine like terms $(+65i - 30i)$ and simplify:

$15 + 65i - 30i - 130i^2 = 15 + 35i - 130i^2$, $i^2 = -1$, then: $15 + 35i - 130i^2 = 15 + 35i - 130(-1) = 15 + 35i + 130 = 145 + 35i$

8) The answer is 28.

Use the Pythagorean theorem to find the value of y: $a^2 + b^2 = c^2 \rightarrow 3^2 + 4^2 = y^2 \rightarrow$

$y^2 = 9 + 16 = 25 \rightarrow y = 5$. The perimeter of the trapezoid is $8 + 3 + 5 + 8 + 4 = 28$ (notice that $x = 4$).

9) The answer is 20.

Solve the system of equations by elimination method.

$\begin{array}{r} 3x - 4y = -20 \\ -x + 2y = 20 \end{array}$ $\rightarrow$ Multiply the second equation by 3, then add it to the first equation.

$\begin{array}{r} 3x - 4y = -20 \\ 3(-x + 2y = 20) \end{array} \Rightarrow \begin{array}{r} 3x - 4y = -20 \\ -3x + 6y = 60 \end{array} \Rightarrow$ adding the equations: $2y = 40 \Rightarrow y = 20$

10) The answer is 7.2 hours.

Use distance formula: $Distance = Rate \times time \Rightarrow 396 = 55 \times T$

Divide both sides by 55. $396 \div 55 = T \Rightarrow T = 7.2\ hours$. Change hours to minutes for the decimal part. $0.2\ hours = 0.2 \times 60 = 12\ minutes$. The answer is 7 hours and 12 minutes or 7.2 hours.

11) The answer is 102.

50% of 60 equals to: $0.50 \times 60 = 30$, 12% of 600 equals to: $0.12 \times 600 = 72$

50% of 60 added to 12% of 600: $30 + 72 = 102$

12) The answer is $(-\infty, -2) \cup (8, +\infty)$.

Factor the numerator: $\frac{6x+12}{x-8} > 0 \rightarrow \frac{6(x+2)}{x-8} > 0$.

Divide both sides by 6: $\frac{\frac{6(x+2)}{x-8}}{6} > \frac{0}{6} \rightarrow \frac{x+2}{x-8} > 0$.

Determine the signs of the factors $\frac{x+2}{x-8}$.

Find the roots of the numerator and denominator:

$x + 2 = 0 \rightarrow x = -2$ and $x - 8 = 0 \rightarrow x = 8$.

Plug in some values of x into the intervals: $(-\infty, -2)$, $(-2, 8)$, and $(8, +\infty)$. Now, check the solutions. Only $x < -2$, or $x > 8$ work in the inequality. The interval notation: $(-\infty, -2) \cup (8, +\infty)$.

13) The answer is $\frac{17}{18}$.

If 17 balls are removed from the bag at random, there will be one ball in the bag. The probability of choosing a brown ball is 1 out of 18. Therefore, the probability of not choosing a brown ball is

17 out of 18 (or $\frac{17}{18}$) and the probability of having not a brown ball after removing 17 balls is the same.

14) The answer is 6.

The minimum value of the function corresponds to the y −coordinate of the point on the graph that has the smallest y −coordinate on the graph. Since the smallest y −coordinate belongs to the point with coordinates $(6, -3)$, the minimum value of the graph is $f(6) = -3$. Therefore, the minimum value of $f(x)$ is at $x = 6$.

15) The answer is $\frac{\sqrt{2}}{2}$.

The value of $sin\ 45° = \frac{\sqrt{2}}{2}$

16) The answer is $2x^3 + x^2 + 2y^5 - 2y^2 + 8z^3$.

To simplify this polynomial, combine like terms. Then:

$$2x^2 + 4y^5 - x^2 + 2z^3 - 2y^2 + 2x^3 - 2y^5 + 6z^3$$
$$= 2x^2 - x^2 + 2x^3 - 2y^2 + 4y^5 - 2y^5 + 2z^3 + 6z^3$$
$$= x^2 + 2x^3 - 2y^2 + 2y^5 + 8z^3 = 2y^5 + 2x^3 + 8z^3 + x^2 - 2y^2$$

Writing the polynomial in standard form:

$$2y^5 + 2x^3 + 8z^3 + x^2 - 2y^2 = 2x^3 + x^2 + 2y^5 - 2y^2 + 8z^3$$

17) The answer is $120x + 14,000 \leq 20,000$.

Let x be the number of new shoes the team can purchase. Therefore, the team can purchase $120x$. The team had \$20,000 and spent \$14,000. Now the team can spend on new shoes \$6,000 at most. Now, write the inequality: $120x + 14{,}000 \leq 20{,}000$.

18) The answer is 125%.

The question is this: 1.75 is what percent of 1.40? Use the percent formula:

$$part = percent \times whole,\ 1.75 = x \times 1.40 \Rightarrow percent = \frac{1.75}{1.40} = 125$$

1.75 is 125% of 1.40.

19) The answer is $48\ cm$.

$One\ liter = 1{,}000\ cm^3 \rightarrow\ 6\ liters = 6{,}000\ cm^3$;

Formula for the volume of a rectangle prism is: $V = l \times w \times h$

$6{,}000 = 25 \times 5 \times h \rightarrow h = \frac{6{,}000}{125} = 48\ cm$

20) The answer is 48.

Red and blue balls are in ration of $2:3$. Write a proportion and solve.

$$\frac{2}{3} = \frac{x}{72} \rightarrow 3x = 2 \times 72 \rightarrow 3x = 144 \rightarrow x = \frac{144}{3} = 48$$

21) The answer is $p = 4$ and $r = 8$.

$(x + 2)(x + p) = x^2 + (2 + p)x + 2p \rightarrow 2 + p = 6 \rightarrow p = 4$ and $r = 2p = 8$

22) The answer is $\frac{7}{5}$.

First, find a common denominator for 2 and $\frac{3x}{x-5}$. It's $x - 5$. Then:

$2 + \frac{3x}{x-5} = \frac{2(x-5)}{x-5} + \frac{3x}{x-5} = \frac{2x-10+3x}{x-5} = \frac{5x-10}{x-5}$.

Now, multiply the numerator and denominator of $\frac{3}{5-x}$ by -1. Then: $\frac{3\times(-1)}{(5-x)\times(-1)} = \frac{-3}{x-5}$. Rewrite the expression: $\frac{5x-10}{x-5} = \frac{-3}{x-5}$. Since the denominators of both fractions are equal, then, the numerators must be equal:

$5x - 10 = -3 \rightarrow 5x = 7 \rightarrow x = \frac{7}{5}$.

23) The answer is $\frac{7x-5}{2x^2-4x}$.

$$\left(\frac{f}{g}\right)(x) = \frac{f(x)}{g(x)} = \frac{7x-5}{2x^2-4x}$$

24) The answer is $\frac{12}{13}$.

$tan\ \theta = \frac{opposite}{adjacent}$, $tan\ \theta = \frac{5}{12} \Rightarrow$ We have the following right triangle. Then:

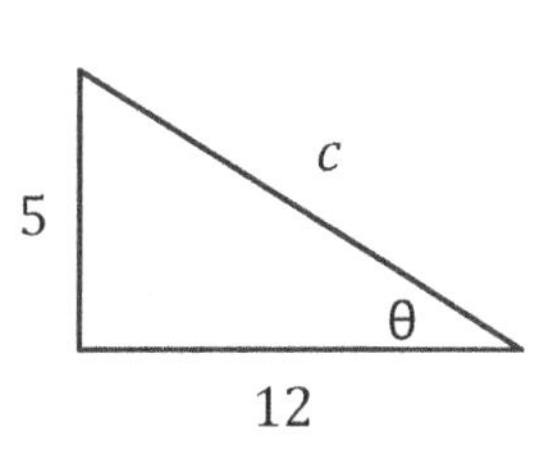

$c = \sqrt{5^2 + 12^2} = \sqrt{25 + 144} = \sqrt{169} = 13$

$$\cos\theta = \frac{adjacent}{hypotenuse} = \frac{12}{13}$$

25) The answer is $x = -\frac{4}{3}$.

To find the vertical asymptote(s) of a rational function, set the denominator equal to 0 and solve for x. Then: $3x + 4 = 0 \rightarrow 3x = -4 \rightarrow x = -\frac{4}{3}$

The vertical asymptote is $x = -\frac{4}{3}$

26) The answer is 100 miles.

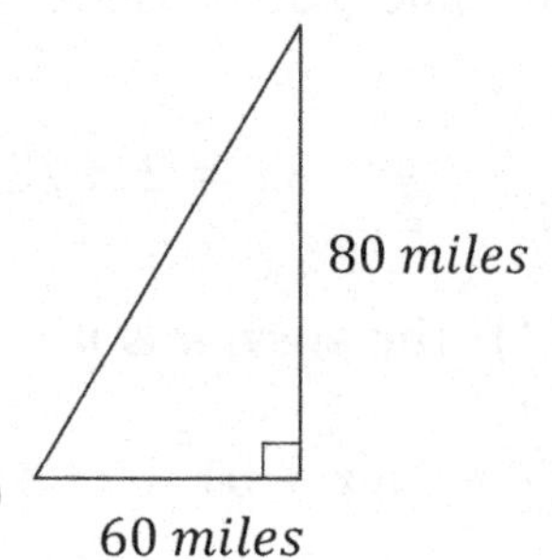

Use the information provided in the question to draw the shape.

Use Pythagorean Theorem: $a^2 + b^2 = c^2$

$60^2 + 80^2 = c^2 \Rightarrow 3{,}600 + 6{,}400 = c^2 \Rightarrow 10{,}000 = c^2 \Rightarrow c = 100$

The boat is 100 miles from its starting point.

27) The answer is 84.

To solve absolute value equations, write two equations. $x - 10$ could be positive 4, or negative 4. Therefore, $x - 10 = 4 \Rightarrow x = 14$, $x - 10 = -4 \Rightarrow x = 6$. Find the product of solutions: $6 \times 14 = 84$

28) The answer is $\frac{9}{20}$.

Set of numbers that are not composite between 1 and 20: $A = \{1, 2, 3, 5, 7, 11, 13, 17, 19\}$

$$Probability = \frac{number\ of\ desired\ outcomes}{number\ of\ total\ outcomes} = \frac{9}{20}$$

29) The answer is $a_n = 5 \times (-4)^{(n-1)}$.

The sequence is a geometric sequence with a common ratio of -4, so the nth term can be found using the formula: $a_n = a_1 r^{(n-1)}$. Substituting the given values, we get:

$a_n = 5 \times (-4)^{(n-1)}$

Therefore, the expression that can be used to find the nth term in the sequence is $a_n = 5 \times (-4)^{(n-1)}$.

30) The answer is 224.

$y = 4ab + 3b^3$. Plug in the values of a and b in the equation: $a = 2$ and $b = 4$

$y = 4(2)(4) + 3(4)^3 = 32 + 3(64) = 32 + 192 = 224$

31) The answer is 12.

The area of a trapezoid can be determined using the formula: $A = \frac{1}{2} \times (a + b) \times h$

We know: $DC = 4\ cm$, $AE = 2\ cm$, and $AD = 4\ cm \rightarrow A = \frac{1}{2} \times (4\ cm + 2\ cm) \times 4\ cm = 12\ cm^2$.

32) The answer is "*no solution*"

When the logarithms have the same base: $f(x) = g(y)$, then: $x = y, log(5x + 2) = log(3x - 1) \rightarrow 5x + 2 = 3x - 1 \rightarrow 5x + 2 - 3x + 1 - 0 \rightarrow 2x + 3 = 0 \rightarrow 2x = -3 \rightarrow x = -\frac{3}{2}$

Verify Solution: $log(5x + 2) = log\ (5(-\frac{3}{2}) + 2) = log\ (-5.5)$

Logarithms of negative numbers are not defined. Therefore, there is no solution for this equation.

33) The answer is $8a(a + b)^2$.

Consider the variable part to be $\boldsymbol{a(a+b)^2}$, then this expression has two like terms with coefficients $\boldsymbol{-15}$ and $\boldsymbol{23}$. Then: $\boldsymbol{-15a(a+b)^2 + 23a(a+b)^2 = 8a(a+b)^2}$.

34) The answer is $r = \frac{ln\left(\frac{P}{1.7}\right)}{t}$.

$\boldsymbol{f(x) = g(x)}$, then: $\boldsymbol{log_a f(x) = log_a g(x)}$.

So, $ln\ P = ln\ 1.7e^{rt}$. Use the logarithm rule: $log_a(x \cdot y) = log_a x + log_a y$.

Therefore, we have $ln\ P = ln\ 1.7 + ln\ e^{rt}$.

Then: $log_a x^y = y\ log_a x$ and $log_a a = 1$.

Accordingly: $ln\ P = ln\ 1.7 + rt \cdot ln\ e \rightarrow ln\ P = ln\ 1.7 + rt$.

Rewrite as: $rt = ln\ P - ln\ 1.7$.

$log_a x - log_a y = log_a \frac{x}{y} \rightarrow rt = ln\left(\frac{P}{1.7}\right)$.

Then divide both sides by t: $r = \frac{ln\left(\frac{P}{1.7}\right)}{t}$.

35) The answer is (Domain: $x \geq -3$, Range: $y \geq 7$).

For domain: Find non-negative values for radicals: $3x + 9 \geq 0$

Domain of functions: $3x + 9 \geq 0 \rightarrow 3x \geq -9 \rightarrow x \geq -3$

Domain of the function $y = 5\sqrt{3x + 9} + 7$: $x \geq -3$

For range: The range of a radical function of the form $c\sqrt{ax + b} + k$ is: $f(x) \geq k$

For the function $y = 5\sqrt{3x + 9} + 7$, the value of k is 7. Then: $y \geq 7$

Range of the function $y = 5\sqrt{3x + 9} + 7$: $y \geq 7$

ALEKS Mathematics Practice Test 7

1) The answer is $\frac{m^{\frac{9}{2}}}{n^{\frac{4}{3}}}$.

$m^{\frac{1}{2}}\, n^{-2}\, m^4\, n^{\frac{2}{3}}$ simplifies to $m^{\frac{1}{2}} \cdot m^4 = m^{\frac{1}{2}+4} = m^{\frac{9}{2}}$, and $n^{-2} \cdot n^{\frac{2}{3}} = n^{-2+\frac{2}{3}} = n^{-\frac{4}{3}} = \frac{1}{n^{\frac{4}{3}}}$, so

$m^{\frac{9}{2}} \cdot \frac{1}{n^{\frac{4}{3}}} = \frac{m^{\frac{9}{2}}}{n^{\frac{4}{3}}}$.

2) The answer is 2.5.

Use the distance formula: $Distance = Rate \times time$

We can write this equation for the distance traveled by the two trains:

$r_1 t + r_2 t = d$, by substituting the speed (rate) of each train, we have:

$95t + 85t = 180$, now solve for t (time). Then: $180t = 450 \rightarrow \frac{180t}{180} = \frac{450}{180} \rightarrow t = 2.5$

After 2.5 hours (two hours and thirty minutes) the two trains will be exactly 450 miles apart.

3) The answer is -5.

Simplify, $\frac{2x^2}{5} - 10 = 0$. First, multiply both sides of the equation by 5:

$5 \times \left(\frac{2x^2}{5} - 10\right) = 5 \times 0 \rightarrow 5 \times \frac{2x^2}{5} - 5 \times 10 = 0 \rightarrow 2x^2 - 50 = 0$.

Add 50 to both sides: $2x^2 - 50 + 50 = 0 + 50 \rightarrow 2x^2 = 50$. Now, divide both sides by 2. Then, $2x^2 \div 2 = 50 \div 2 \rightarrow x^2 = 25$. Therefore, $x = \pm 5$.

4) The answer is $(4, 3)$.

To solve this system of equations, add the two equations. Then:

$$\begin{cases} x + 2y = 10 \\ 6x - 2y = 18 \end{cases} \rightarrow x + 6x + 2y - 2y = 10 + 18 \rightarrow 7x = 28 \rightarrow x = 4$$

Substitute the value of x in the first equation and solve for y:

$x + 2y = 10, x = 4 \rightarrow 4 + 2y = 10 \rightarrow 2y = 10 - 4 \rightarrow 2y = 6 \rightarrow y = 3$

5) The answer is $x = 1$.

The function $g(x) = -\frac{1}{8}(x-1)^2 - 3$ is quadratic and in the vertex form $y = a(x-h)^2 + k$. So, there is an axis of symmetry parallel to the y −axis that passes through the vertex. The vertex is the point with the coordinate $(h, k) = (1, -3)$. Therefore, the axis of symmetry is $x = 1$.

6) The answer is 6.

To find the value of c that makes the graph of h 12 units above the graph of f, we can set up an equation: $h(x) = f(x) + 12$.

Substitute the given expressions for $f(x)$ and $h(x)$ into this equation:

$3x^2 + c = 3x^2 - 6 + 12$.

Simplify and solve for c: $c = 6$.

Therefore, the value of c that will make the graph of h 12 units above the graph of f is 6.

7) The answer is $\$770$.

Use simple interest formula: $I = prt$ (I = interest, p = principal, r = rate, t = time)

$I = (11{,}000)(0.035)(2) = 770$

If you deposit $11,000 in the bank, you will earn $770 in interest in two years.

8) The answer is 9, 600.

Number of visiting fans: $\frac{2\times 24{,}000}{5} = 9{,}600$ (notice that the ratio of visiting fans to total seats is 2 to 5)

9) The answer is $120°$.

The sum of all angles in a quadrilateral is 360 degrees. Let x be the smallest angle in the quadrilateral. Then the angles are: $2x, 3x, 3x, 4x$, so:

$2x + 3x + 3x + 4x = 360 \rightarrow 12x = 360 \rightarrow x = 30$

The angles in the quadrilateral are $60°, 90°, 90°$, and $120°$.

10) The answer is 30.

Plug in the value of x in the equation and solve for y. $2y = \frac{2x^2}{3} + 6 \rightarrow 2y = \frac{2(9)^2}{3} + 6 \rightarrow$

$2y = \frac{2(81)}{3} + 6 \rightarrow 2y = 54 + 6 \rightarrow 2y = 60 \rightarrow y = 30.$

11) The answer is $90\ cm$.

The length of the rectangle is $\frac{5}{4} \times 20 = 25\ cm$, perimeter of the rectangle is:

$2 \times (length + width) = 2 \times (20 + 25) = 90\ cm$

12) The answer is 70.

The perimeter of the trapezoid is $36\ cm$. Therefore, the missing side (height) is:

$36 - 8 - 12 - 6 = 10.$

Area of a trapezoid: $A = \frac{1}{2} h\ (b_1 + b_2) = \frac{1}{2}(10)(6 + 8) = 70.$

13) The answer is $18°$.

The sum of supplement angles is 180. Let x be that angle. Therefore, $x + 9x = 180$

$10x = 180$, divide both sides by 10: $x = 18$

The measure of the angle is 18 degrees.

14) The answer is $x = \frac{3}{2}$.

Use the cross multiplication to solve for x: $\frac{5}{x+1} = \frac{x+1}{x^2-1} \rightarrow 5(x^2 - 1) = (x + 1)(x + 1)$

Simplify $5(x^2 - 1)$ using the distributive property: $5(x^2 - 1) = 5x^2 - 5$

Simplify $(x + 1)(x + 1)$ using the FOIL (First-Out-In-Last) method:

$(x + 1)(x + 1) = x^2 + x + x + 1 = x^2 + 2x + 1$

Then: $5(x^2 - 1) = (x + 1)(x + 1) \rightarrow 5x^2 - 5 = x^2 + 2x + 1$. Since this is a quadratic equation, we need to bring all terms to one side of the equation. Subtract $(x^2 + 2x + 1)$ from both sides. Then:

$5x^2 - 5 - (x^2 + 2x + 1) = x^2 + 2x + 1 - (x^2 + 2x + 1)$

Simplify and combine like terms: $5x^2 - 5 - x^2 - 2x - 1 = x^2 + 2x + 1 - (x^2 + 2x + 1) \rightarrow$

$4x^2 - 2x - 6 = 0$

Use the quadratic formula to solve for x. Then: $x_{1,2} = \frac{-b \pm \sqrt{b^2-4ac}}{2a} = \frac{-(-2) \pm \sqrt{(-2)^2-4(4)(-6)}}{2(4)} =$

$\frac{-(-2) \pm \sqrt{100}}{8} \rightarrow x = \frac{-(-2)+10}{8} = \frac{2+10}{8} = \frac{12}{8} = \frac{3}{2}$ or $x = \frac{-(-2)-10}{8} = \frac{2-10}{8} = \frac{-8}{8} = -1$

The solution $x = -1$ is not defined in the original equation. (If $x = -1$, then the denominators are equal to zero, which is not defined.) Therefore, the solution $x = \frac{3}{2}$ is the only accepted solution.

15) The answer is 60.

First, find the value of b, and then find $f(3)$. Since $f(2) = 35$, substuting 2 for x and 35 for $f(x)$ gives $35 = b(2)^2 + 15 \rightarrow 35 = 4b + 15$. Solving this equation gives $b = 5$. Thus
$f(x) = 5x^2 + 15, \quad f(3) = 5(3)^2 + 15 \rightarrow f(3) = 45 + 15 \rightarrow f(3) = 60$.

16) The answer is $\frac{\sqrt{8}}{3}$.

$sin\, A = \frac{1}{3} \Rightarrow$ Since $sin\, \theta = \frac{opposite}{hypotenuse}$, we have the following right triangle. Then:

3
1
A
c

$c = \sqrt{3^2 - 1^2} = \sqrt{9-1} = \sqrt{8}$, $cos\, \theta = \frac{adjacent}{hypotenuse} \Rightarrow cos\, A = \frac{\sqrt{8}}{3}$

17) The answer is $25\,\pi$.

The equation of a circle in standard form is $(x-h)^2 + (y-k)^2 = r^2$, where r is the radius of the circle. In this circle, the radius is 5. $r^2 = 25 \rightarrow r = 5$, $(x+2)^2 + (y-4)^2 = 25$

Area of a circle: $A = \pi r^2 = \pi(5)^2 = 25\,\pi$

18) The answer is $x = 7$.

$log_3(x+20) - log_3(x+2) = 1$. First, condense the two logarithms:

$log_3(x+20) - log_3(x+2) = 1 \rightarrow log_3\left(\frac{x+20}{x+2}\right) = 1$.

We know that: $log_a\, a = 1$. Then:

$log_3\left(\frac{x+20}{x+2}\right) = 1 \rightarrow log_3\left(\frac{x+20}{x+2}\right) = log_3 3 \rightarrow \frac{x+20}{x+2} = 3$.

Use cross multiplication and solve for x.

$\frac{x+20}{x+2} = 3 \rightarrow x + 20 = 3(x + 2) \rightarrow x + 20 = 3x + 6 \rightarrow 2x = 14 \rightarrow x = 7.$

19) The answer is -36.

The standard form equation of a parabola is $y = ax^2 + bx + c$.

$y = (x + 3)(x - 9) \rightarrow y = x^2 - 6x - 27.$

So, the x –coordinate of the vertex is $\frac{-b}{2a} = \frac{-(-6)}{2(1)} = 3.$

Substituting 3 in the original equation to get the y –coordinate, we get:

$y = 3^2 - 6(3) - 27 = -36.$

20) The answer is 40.

First, find the number. Let x be the number. Write the equation and solve for x.

150% of a number is 75, then: $1.5 \times x = 75 \Rightarrow x = \frac{75}{1.5} = 50$

80% of 50 is: $0.8 \times 50 = 40$

21) The answer is $c(3, -2)$, radius $= 3$.

The standard form of the circle equation is: $(x - h)^2 + (y - k)^2 = r^2$ where the radius of the circle is r, and it's centered at (h, k).

First, move the loose number to the right side: $x^2 + y^2 - 6x + 4y = -4$

Group x-variables and y-variables together: $(x^2 - 6x) + (y^2 + 4y) = -4$

Convert x to square form:

$(x^2 - 6x + 9) + y^2 - 6y = -4 + 9 \rightarrow (x - 3)^2 + (y^2 + 4y) = -4 + 9$

Convert y to square form:

$(x - 3)^2 + (y^2 + 4y + 4) = -4 + 9 + 4 \rightarrow (x - 3)^2 + (y + 2)^2 = 9$

Then, the equation of the circle in standard form is: $(x - 3)^2 + (y + 2)^2 = 3^2$

The center of the circle is at $(3, -2)$ and its radius is 3.

22) The answer is 2.

Let the equation be $\boldsymbol{P(x) = 6x^2 + 4x - 10}$. Factor $\mathbf{2}$ from the equation: $\boldsymbol{6x^2 + 4x - 10 = 2(3x^2 + 2x - 5)}$. We just need to factor the expression $\boldsymbol{Q(x) = 3x^2 + 2x - 5}$. First, find the

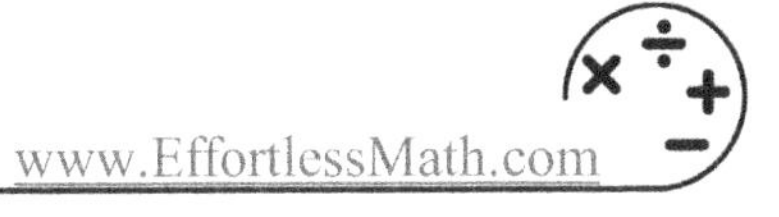

roots of the equation $3x^2 + 2x - 5 = 0$ by evaluating the discriminant expression $\Delta = b^2 - 4ac$ of the quadratic equation as $ax^2 + bx + c = 0$. So, we have:

$\Delta = (2)^2 - 4(3)(-5) = 4 + 60 = 64 \rightarrow \Delta > 0$

Use the quadratic formula: $x_{1,2} = \frac{-b \pm \sqrt{\Delta}}{2a}$. Therefore, the zeros of the equation $3x^2 + 2x - 5 = 0$ are $x_1 = \frac{-2+\sqrt{64}}{2(3)} = \frac{-2+8}{6} = 1$ or $x_2 = \frac{-2-\sqrt{64}}{2(3)} = \frac{-2-8}{6} = -\frac{10}{6} = -\frac{5}{3}$.

Then, the equation $Q(x) = 0$, can be written in factored form as

$(x - 1)\left(x + \frac{5}{3}\right) = 0$

Multiply both sides of the equation by 3:

$3 \times (x - 1)\left(x + \frac{5}{3}\right) = 3 \times 0 \rightarrow (x - 1)(3x + 5) = 0$

That is equivalent to the equation $3x^2 + 2x - 5 = 0$. Now, to obtain the expression $P(x)$, multiply both sides of the expression $Q(x)$ by 2. Thus,

$P(x) = 2Q(x) \rightarrow P(x) = 2(x - 1)(3x + 5) \rightarrow P(x) = (2x - 2)(3x + 5)$

Compare the resulting factor with the factored expression in the content of the question:

$(2x - 2)(3x + 5) = (2x - m)(3x + 5)$

You can see that the value of m is 2.

23) The answer is $\frac{8}{17}$.

$\tan\theta = \frac{opposite}{adjacent}$, and $\tan x = \frac{8}{15}$, therefore, the opposite side of the angle x is 8 and the adjacent side is 15. Let's draw the triangle.

c
8
x°
15

Using Pythagorean theorem, we have:

$a^2 + b^2 = c^2 \rightarrow 8^2 + 15^2 = c^2 \rightarrow 64 + 225 = c^2 \rightarrow c = 17$, $\sin x = \frac{opposite}{hypotenuse} = \frac{8}{17}$

24) The answer is $x^{\frac{63}{8}}$.

Use Exponent's rules: $(x^a)^b = x^{a \times b}$. Then: $(x^7)^{\frac{9}{8}} = x^{7 \times \frac{9}{8}} = x^{\frac{63}{8}}$

25) The answers are $f(x) = x^3 - 3x^2 - x + 3$.

-1, 1 and 3 are the roots of the function so: $x = 1 \rightarrow x - 1 = 0$.

$x = -1 \rightarrow x + 1 = 0$

$x = 3 \rightarrow x - 3 = 0$

Multiply the terms together: $f(x) = (x-1)(x+1)(x-3) \rightarrow f(x) = (x^2-1)(x-3)$.

FOIL: $f(x) = (x^2-1)(x-3) \rightarrow f(x) = x^3 - 3x^2 - x + 3$.

26) The answer is 2.

We know that: $sin^2a + cos^2a = 1$

Then: $x + sin^2a + cos^2a = 3 \rightarrow x + 1 = 3 \rightarrow x = 2$

27) The answer is $\frac{y}{6}$.

Solve for x. $\sqrt{6x} = \sqrt{y}$. Square both sides of the equation: $\left(\sqrt{6x}\right)^2 = \left(\sqrt{y}\right)^2 \rightarrow$

$6x = y \rightarrow x = \frac{y}{6}$

28) The answer is 62.12.

$average = \frac{sum\ of\ terms}{number\ of\ terms}$, the sum of the weight of all girls is: $18 \times 57 = 1{,}026\ kg$

The sum of the weight of all boys is: $32 \times 65 = 2{,}080\ kg$. The sum of the weight of all students is: $1{,}026 + 2{,}080 = 3{,}106\ kg$. $average = \frac{3{,}106}{50} = 62.12$

29) The answer is 0.

By definition, the x –intercept is the point where a graph crosses the x –axis. If $y = 0$ is put into the equation, it would become:

$y = \frac{x-1}{1-x^2} \rightarrow \frac{x-1}{1-x^2} = 0 \rightarrow x - 1 = 0 \rightarrow x = 1$.

However, we know that the function is not defined for $x = 1$ and $x = -1$, as these values would result in division by zero.

Therefore, the graph of $y = \frac{x-1}{1-x^2}$ would not have x –intercepts.

30) The answer is on the following graph.

A quadratic function in vertex form is: $y = a(x-h)^2 + k$ and (h, k) is the vertex. Then, the vertex of $y = (x+1)^2 - 2$ is $(-1, -2)$.

Substitute zero for x and solve for y: $y = (0+1)^2 - 2 = -1$.

The y-intercept is $(0, -1)$. Now, you can simply graph the quadratic function. Notice that the quadratic function is a U-shaped curve. (You can plug in some values of x and solve for y to get some points on the graph.)

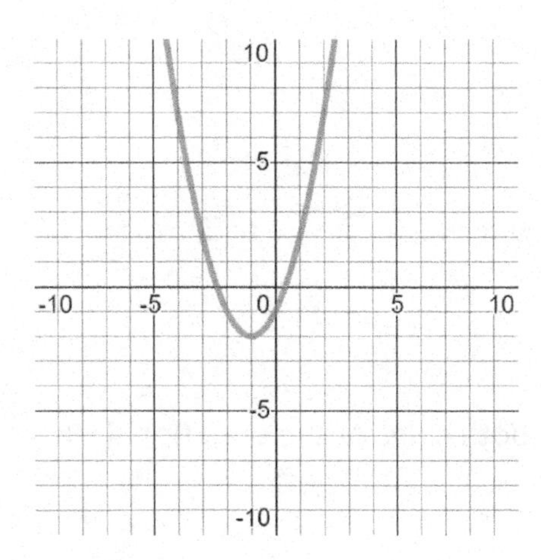

31) The answer is $3\frac{1}{3}$.

First, factorize the numerator and simplify.

$\frac{x^2-9}{x+3} + 2(x+4) = 15 \rightarrow \frac{(x-3)(x+3)}{x+3} + 2x + 8 = 15, \rightarrow x - 3 + 2x + 8 = 15 \rightarrow 3x + 5 = 15$

Subtract 5 from both sides of the equation. Then: $\rightarrow 3x = 15 - 5 = 10 \rightarrow x = \frac{10}{3} = 3\frac{1}{3}$.

32) The answer is 121.

Plug in the value of x and y in the expression. $x = 4$ and $y = 9$

$7x^2 - 2xy + y^2 = 7(4)^2 - 2(4)(9) + (9)^2 = 7(16) - 2(36) + 81 \rightarrow 112 - 72 + 81 = 121$

33) The answer is 1.

The intersection of two functions is the point with 2 for x. Then:

$f(2) = g(2)$ and $g(2) = (2 \times (2)) - 3 = 4 - 3 = 1$

Then, $f(2) = 1 \rightarrow a(2)^2 + b(2) + c = 1 \rightarrow 4a + 2b + c = 1$ (i)

The value of x in the vertex of the parabola is given by: $x = -\frac{b}{2a} \rightarrow -2 = -\frac{b}{2a} \rightarrow b = 4a$ (ii)

In the point $(-2, 5)$, the value of the $f(x)$ is 5.

$f(-2) = 5 \rightarrow a(-2)^2 + b(-2) + c = 5 \rightarrow 4a - 2b + c = 5$ (iii)

Using the first two equations:

$\begin{cases} 4a+2b+c=1 \\ 4a-2b+c=5 \end{cases} \rightarrow$

Equation 1 minus equation 2 is: (i)−(iii) $\rightarrow 4b=-4 \rightarrow b=-1$ (iv)

Plug in the value of b in the second equation: $b=4a \quad \rightarrow a=\frac{b}{4}=-\frac{1}{4}$

Plug in the values of a and b in the first equation. Then:

$\rightarrow 4\left(\frac{-1}{4}\right)+2(-1)+c=1 \rightarrow -1-2+c=1 \rightarrow c=1+3 \rightarrow c=4$

The product of a, b, and $c=\left(-\frac{1}{4}\right)\times(-1)\times 4=1$.

34) The answer is 15.

Use fractions division rule: $\frac{a}{b}\div\frac{c}{d}=\frac{a}{b}\times\frac{d}{c}=\frac{a\times d}{b\times c}$

Then: $\frac{5x}{x+2}\div\frac{x}{3x+6}=\frac{5x}{x+2}\times\frac{3x+6}{x}=\frac{5x(3x+6)}{x(x+2)}=\frac{5x\times3(x+2)}{x(x+2)}$

Cancel common factor: $\frac{5x\times3(x+2)}{x(x+2)}=\frac{15x(x+2)}{x(x+2)}=15$

35) The answer is $x=\frac{ln\,(12)}{2}$.

If $f(x)=g(x), then: ln(f(x))=ln(g(x)) \rightarrow ln(e^{2x})=ln(12)$

Use logarithm rule: $log_a x^b=b\,log_a x \rightarrow ln(e^{2x})=2x\,ln(e) \rightarrow (2x)ln(e)=ln(12)$

$ln(e)=1$, then: $(2x)ln(e)=ln(12) \rightarrow 2x=ln(12) \rightarrow x=\frac{ln\,(12)}{2}$

ALEKS Mathematics Practice Test 8

1) The answer is 2.

There can be 0, 1, or 2 solutions to a quadratic equation. In standard form, a quadratic equation is written as $ax^2 + bx + c = 0$.

For the quadratic equation, the expression $b^2 - 4ac$ is called the discriminant. If the discriminant is positive, there are 2 distinct solutions for the quadratic equation. If the discriminant is 0, there is one solution for the quadratic equation and if it is negative the equation does not have any solutions.

To find the number of solutions for $x^2 = 4x - 3$, first, rewrite it as $x^2 - 4x + 3 = 0$.

Find the value of the discriminant. $b^2 - 4ac = (-4)^2 - 4(1)(3) = 16 - 12 = 4$.

Since the discriminant is positive, the quadratic equation has two distinct solutions.

2) The answers are -512 and $\frac{1}{625}$.

$(-8)^3 = (-8)(-8)(-8) = -512$

$$5^{-4} = \frac{1}{(5)(5)(5)(5)} = \frac{1}{625}$$

3) The answer is 29.

Here we can substitute 8 for x in the equation. Thus, $y - 3 = 2(8 + 5)$, $y - 3 = 26$

Adding 3 to both sides of the equation: $y = 26 + 3 \rightarrow y = 29$

4) The answer is -108.

$12x^4 + n = a(x^2 + 3)(x^2 - 3) = ax^4 - 9a \rightarrow a = 12$, and $n = -9a = -9 \times 12 = -108$

5) The answer is $(v^2 + 4)(5v - 3)$.

Factor the polynomial by grouping:

$5v^3 - 3v^2 + 20v - 12 = (5v^3 - 3v^2) + (20v - 12)$. Now, factorize each parenthesis. $(5v^3 - 3v^2) + (20v - 12) = v^2(5v - 3) + 4(5v - 3)$. Take the common factor $(5v - 3)$ out. Then: $v^2(5v - 3) + 4(5v - 3) = (v^2 + 4)(5v - 3)$

6) The answer is $-\frac{8}{7}$.

Since $f(x)$ is a linear function with a negative slope, then when $x = -2, f(x)$ is maximum, and when $x = 3, f(x)$ is minimum. Then the ratio of the minimum value to the maximum value of the function is: $\frac{f(3)}{f(-2)} = \frac{-3(3)+1}{-3(-2)+1} = \frac{-8}{7} = -\frac{8}{7}$

7) The answer is 2.

Since $y = f(x)$, the value of $f(0)$ is equal to the value of $f(x)$, or y, when $x = 0$. The graph indicates that when $x = 0, y = 2$. It follows that the value of $f(0) = 2$.

8) The answer is 20.

Given the two equations, substitute the numerical value of a into the second equation to solve for x. $a = \sqrt{5}, 4a = \sqrt{4x}$

Substituting the numerical value for a into the equation with x is as follows.

$4(\sqrt{5}) = \sqrt{4x}$, from here, distribute the 4. $4\sqrt{5} = \sqrt{4x}$

Now square both sides of the equation. $(4\sqrt{5})^2 = (\sqrt{4x})^2$

Remember to square both terms within the parentheses. Also, recall that squaring a square root sign cancels them out. $4^2\sqrt{5}^2 = 4x \rightarrow 16(5) = 4x \rightarrow 80 = 4x \rightarrow x = 20$

9) The answer is $a_n = 4n + 3$.

To find the recursive formula, start by looking at the common differences and ratios of consecutive terms. By evaluating the difference between terms with the previous term, you notice that the differences between consecutive terms are all the same. That is, $a_n - a_{n-1} = 4$. In this step, for calculating the first term of the arithmetic sequence, substitute one of the given terms like $a_4 = 19$ and the common difference $d = 4$ in the arithmetic sequence formula: $a_n = a_1 + d(n-1)$, where $a_1 =$ the first term, $d =$ the common difference between terms, $n =$ number of items. Then,

$a_4 = a_1 + 4(4-1) = 19 \rightarrow a_1 + 12 = 19 \rightarrow a_1 = 7$

Therefore, the nth term of the sequence is $a_n = 7 + 4(n-1)$. Now, simplify as $a_n = 4n + 3$.

10) The answer is 2.2904×10^3.

To multiply two numbers in scientific notation, multiply their coefficients and add their exponents. For these two numbers in scientific notation, multiply the coefficients: $4.09 \times 5.6 = 22.904$

Add the powers of 10: $10^6 \times 10^{-4} = 10^{6+(-4)} = 10^2$. Then:

$$(4.09 \times 10^6) \times (5.6 \times 10^{-4}) = (4.09 \times 5.6) \times (10^6 \times 10^{-4}) = 22.904 \times (10^{6+(-4)})$$
$$= 2.2904 \times 10^3$$

11) The answer is $\frac{(x-5)(x+4)}{(x-5)+(x+4)}$ or $\frac{x^2-x-20}{2x-1}$.

To rewrite $\frac{1}{\frac{1}{x-5}+\frac{1}{x+4}}$, first simplify $\frac{1}{x-5}+\frac{1}{x+4}$.

$$\frac{1}{x-5}+\frac{1}{x+4}=\frac{1(x+4)}{(x-5)(x+4)}+\frac{1(x-5)}{(x+4)(x-5)}=\frac{(x+4)+(x-5)}{(x+4)(x-5)}$$

Then: $\frac{1}{\frac{1}{x-5}+\frac{1}{x+4}}=\frac{1}{\frac{(x+4)+(x-5)}{(x+4)(x-5)}}=\frac{(x-5)(x+4)}{(x-5)+(x+4)}=\frac{x^2-x-20}{2x-1}$. (Remember, $\frac{1}{\frac{1}{x}}=x$)

12) The answer is 750.

The inequalities $y \le -15x + 3{,}000$ and $y \le 5x$ can be graphed in the xy −plane. They are represented by the lower half-planes with the boundary lines $y = -15x + 3{,}000$ and $y = 5x$, respectively. The solution set of the system of inequalities will be the intersection of these half-planes, including the boundary lines, and the solution (a, b) with the greatest possible value of b will be the point of intersection of the boundary lines. The intersection of boundary lines of these inequalities can be found by substituting $5x$ for y in the equation for the first line: $5x = -15x + 3{,}000$, which has solution $x = 150$. Thus, the x −coordinate of the point of intersection is 150. Therefore, the y −coordinate of the point of intersection of the boundary lines is $y = 5(150) = -15(150) + 3{,}000 = 750$. This is the maximum possible value of b for a point $(a, b) = (150, 750)$ that is in the solution set of the system of inequalities.

13) The answer is 40.

The area of ΔBED is 16, then: $\frac{4 \times AB}{2} = 16 \rightarrow 4 \times AB = 32 \rightarrow AB = 8$

The area of ΔBDF is 18, then: $\frac{3\times BC}{2} = 18 \rightarrow 3 \times BC = 36 \rightarrow BC = 12$

The perimeter of the rectangle is = $2 \times (width + length) = 2 \times (8 + 12) = 40$

14) The answer is 11.5.

The y −intercept in a graph is the value that intercepts the y −axis of the function. For this purpose, calculate the value of the function for $x = 0$. Therefore,

$x = 0 \rightarrow f(0) = 11.5\ (0.8)^0 = 11.5 \times 1 = 11.5$

15) The answer is $11\ m$ by $8\ m$.

To solve the problem, we can use the formula for the perimeter of a rectangle, which is: Perimeter $= 2 \times$ Length $+2 \times$ Width.

We are given that the perimeter of the rectangle is 38 centimeters. We are also given that the length can be represented as $(x + 6)$ and the width can be represented as $(2x - 2)$. So, we can substitute these values into the formula for the perimeter and solve for x:

$38 = 2(x + 6) + 2(2x - 2) \rightarrow 38 = 2x + 12 + 4x - 4 \rightarrow 38 = 6x + 8$

$\rightarrow 30 = 6x \rightarrow x = 5.$

Now, we can find the length and width of the rectangle by substituting $x = 5$ into the expressions for the length and width:

Length $= x + 6 = 5 + 6 = 11\ m$, and width $= 2x - 2 = 2(5) - 2 = 8\ m$.

Therefore, the dimensions of the rectangle are $11\ m$ by $8\ m$.

16) The answer is $420°$.

Use this formula: Degrees $=$ Radians $\times \frac{180}{\pi}$

Radians $= \frac{7\pi}{3} \times \frac{180}{\pi} = \frac{1260\pi}{3\pi} = 420°$

17) The answers are $-\frac{5\pi}{2}$ and $\frac{7\pi}{2}$.

Coterminal angles are equal angles. To find a coterminal of an angle, add or subtract 360 degrees (or 2π for radians) to the given angle. Then:

Positive angle: $\frac{\pi}{3} + 2\pi = \frac{7\pi}{2}$

Negative angle: $\frac{\pi}{3} - 2\pi = -\frac{5\pi}{2}$

18) The answer is $4\sqrt{3}$.

Based on triangle similarity theorem: $\frac{a}{a+b} = \frac{c}{3} \rightarrow c = \frac{3a}{a+b} = \frac{3\sqrt{3}}{3\sqrt{3}} = 1 \rightarrow$ The area of shaded region is: $\left(\frac{c+3}{2}\right)(b) = \frac{4}{2} \times 2\sqrt{3} = 4\sqrt{3}$

The area of the shaded region is $4\sqrt{3}$ square units.

19) The answer is 5.

The formula for the surface area of a cylinder is:

$SA = 2\pi r^2 + 2\pi rh \rightarrow 150\pi = 2\pi r^2 + 2\pi r(10)$

Both sides divided by 2π: $\rightarrow r^2 + 10r - 75 = 0$

$(r + 15)(r - 5) = 0 \rightarrow r = 5$ or $r = -15$ (unacceptable)

20) The answer is $16.67\ m$.

The rate of construction company $= \frac{25\ cm}{1\ min} = 25\ \frac{cm}{min}$

The height of the wall after 50 minutes $= \frac{25\ cm}{1\ min} \times 50\ min = 1{,}250\ cm$

Let x be the height of wall, then $\frac{3}{4}x = 1{,}250\ cm \rightarrow x = \frac{4 \times 1{,}250}{3} \rightarrow x = 1{,}666.67\ cm = 16.67\ m$

21) The answer is $\frac{4}{7}$.

$\frac{8+3x}{7} - \frac{4-4x}{7} = \frac{8+3x-4+4x}{7} = \frac{7x+4}{7} = \frac{7x}{7} + \frac{4}{7} = x + \frac{4}{7}$. Therefore, this expression is $\frac{4}{7}$ greater than x.

22) The answer is $12\ m$.

The width of the rectangle is twice its length. Let x be the length. Then, $width = 2x$

The perimeter of the rectangle is $2\ (width + length) = 2(2x + x) = 72 \Rightarrow 6x = 72 \Rightarrow x = 12$. The length of the rectangle's length is 12 meters.

23) The answer is $\frac{1}{50}$.

Write the ratio of $5a$ to $2b$. $\frac{5a}{2b} = \frac{1}{20}$. Use cross multiplication and then simplify.

$5a \times 20 = 2b \times 1 \rightarrow 100a = 2b \rightarrow a = \frac{2b}{100} = \frac{b}{50}$

Now, find the ratio of a to b. $\frac{a}{b} = \frac{\frac{b}{50}}{b} \rightarrow \frac{b}{50} \div b = \frac{b}{50} \times \frac{1}{b} = \frac{b}{50b} = \frac{1}{50}$

The ratio of a to b is 1 to 50 or $\frac{1}{50}$.

24) The answer is $\frac{3}{5}$.

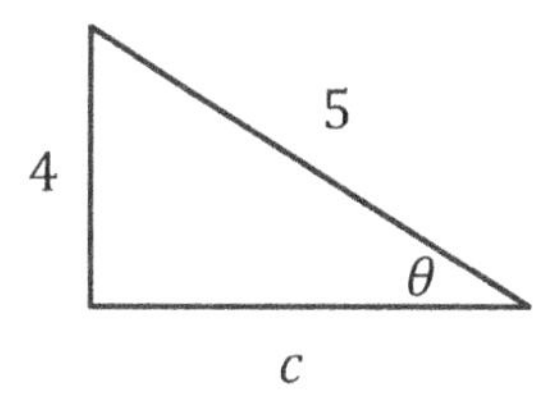

$sin\ \theta = \frac{opposite}{hypotenuse} = \frac{4}{5} \Rightarrow$ We have the following triangle.

Then: $c = \sqrt{5^2 - 4^2} = \sqrt{25 - 16} = \sqrt{9} = 3$, $cos\ \theta = \frac{adjacent}{hypotenuse} = \frac{3}{5}$

25) The answer is 225.

$0.6x = (0.3) \times 20 \rightarrow x = \frac{6}{0.6} = 10 \rightarrow (x + 5)^2 = (15)^2 = 225$

26) The answer is $\frac{3}{2}$.

To solve for the inverse function, first, replace $f(x)$ with y. Then, solve the equation for x and after that, replace every x with y and replace every y with x. Finally, replace y with $f^{-1}(x)$.

$f(x) = \frac{10x - 3}{6} \Rightarrow y = \frac{10x - 3}{6} \Rightarrow 6y = 10x - 3 \Rightarrow 6y + 3 = 10x \Rightarrow \frac{6y + 3}{10} = x$

$f^{-1}(x) = \frac{6x + 3}{10} \Rightarrow f^{-1}(2) = \frac{6(2) + 3}{10} = \frac{15}{10} = \frac{3}{2}$

27) The answer is $\frac{10}{3}$.

METHOD ONE

$log_4(x + 2) - log_4(x - 2) = 1$, Add $log_4(x - 2)$ to both sides

$log_4(x + 2) - log_4(x - 2) + log_4(x - 2) = 1 + log_4(x - 2)$

$log_4(x + 2) = 1 + log_4(x - 2)$

Apply logarithm rule: $a = log_b(b^a) \Rightarrow 1 = log_4(4^1) = log_4(4)$

Then: $log_4(x + 2) = log_4(4) + log_4(x - 2)$

Logarithm rule: $log_c(a) + log_c(b) = log_c(ab)$

Then: $log_4(4) + log_4(x-2) = log_4(4(x-2))$

$log_4(x+2) = log_4(4(x-2))$

When the logs have the same base: $log_b(f(x)) = log_b(g(x)) = f(x) = g(x)$

$(x+2) = 4(x-2), x = \frac{10}{3}$

METHOD TWO

We know that: $log_a b - log_a c = log_a \frac{b}{c}$ and $log_a b = c \Rightarrow b = a^c$

Then: $log_4(x+2) - log_4(x-2) = log_4 \frac{x+2}{x-2} = 1 \Rightarrow \frac{x+2}{x-2} = 4^1 = 4 \Rightarrow x+2 = 4(x-2)$

$\Rightarrow x+2 = 4x-8 \Rightarrow 4x - x = 8+2 \rightarrow 3x = 10 \Rightarrow x = \frac{10}{3}$

28) The answer is 49.

Since $4x = \frac{48}{3}$, and 48 divided by 3 is 16, which gives equation $4x = 16$, then dividing both sides of $4x = 16$ by 4 gives $x = 4$. Therefore, $x - 2 = 4 - 2 = 2$, and 7 to the power of 2 is 49.

29) The answer is $16\sqrt{3}\ cm^2$.

The area of the triangle is: $\frac{1}{2}\ AD \times BC$ and AD is perpendicular to BC. Triangle ADC is a $30° - 60° - 90°$ right triangle. The relationship among all sides of right triangle $30° - 60° - 90°$ is provided in the following triangle: in this triangle, the opposite side of $30°$ angle is half of the hypotenuse. And the opposite side of $60°$ is opposite of $30° \times \sqrt{3}$

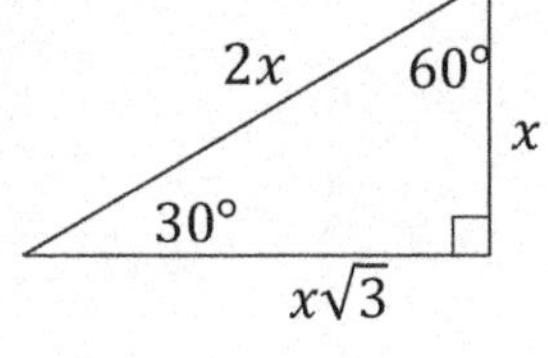

$CD = 4$, then: $AD = 4 \times \sqrt{3}$

The area of the triangle ABC is: $\frac{1}{2}\ AD \times BC = \frac{1}{2}\ 4\sqrt{3} \times 8 = 16\sqrt{3}\ cm^2$

30) The answer is -8.

The problem asks for the sum of the roots of the quadratic equation $2n^2 + 16n + 24 = 0$. Dividing each side of the equation by 2 gives $n^2 + 8n + 12 = 0$. If the roots of $n^2 + 8n + 12 = 0$ are n_1 and n_2, then the equation can be factored as $n^2 + 8n + 12 = (n - n_1)(n - n_2) = 0$.

Looking at the coefficient of n on each side of $n^2 + 8n + 12 = (n + 6)(n + 2)$ gives $n = -6$ or $n = -2$, then, $-6 + (-2) = -8$.

31) The answer is $\frac{u^{10}}{96\,v^6\,w}$.

First, simplify the numerator by using the exponent's rules: $(x^a)^b = x^{a\times b}$. Then:

$\frac{(4u^{-5}v^3)^{-2}}{6w} = \frac{4^{-2}u^{10}v^{-6}}{6w}$. Now, use negative exponent's rule: $\left(\frac{x^a}{x^b}\right)^{-2} = \left(\frac{x^b}{x^a}\right)^2$

Then: $\frac{4^{-2}u^{10}v^{-6}}{6w} = \frac{u^{10}}{(4^2)(6w)(v^6)} = \frac{u^{10}}{96\,v^6\,w}$

32) The answer is 4.

Plug in the values of x and y in the equation of the parabola. Then:

$12 = a(2)^2 + 5(2) + 10 \rightarrow 12 = 4a + 10 + 10 \rightarrow 12 = 4a + 20$

$\rightarrow 4a = 12 - 20 = -8 \rightarrow a = \frac{-8}{4} = -2 \rightarrow a^2 = (-2)^2 = 4.$

33) The answer is $2(\sqrt{12} + 3)$.

Multiply by the conjugate: $\frac{\sqrt{12}+3}{\sqrt{12}+3} \rightarrow \frac{6}{\sqrt{12}-3} \times \frac{\sqrt{12}+3}{\sqrt{12}+3}$.

$(\sqrt{12} - 3)(\sqrt{12} + 3) = 3$, then: $\frac{6}{\sqrt{12}-3} \times \frac{\sqrt{12}+3}{\sqrt{12}+3} = \frac{6(\sqrt{12}+3)}{3} = 2(\sqrt{12} + 3)$.

34) The answer is $-\frac{3}{113} - \frac{45}{113}i$.

To rationalize this imaginary expression, multiply both the numerator and denominator by the conjugate $\frac{15-i}{15-i}$. Then: $\frac{-6i}{15+i} = \frac{-6i\,(15-i)}{(15+i)(15-i)} =$

Apply complex arithmetic rule: $(a + bi)(a - bi) = a^2 + b^2$. Then:

$$\frac{-6i(15 - i)}{(15 + i)(15 - i)} = \frac{-90i + 6i^2}{15^2 - i^2} = \frac{-90i - 6}{226} = -\frac{6}{226} - \frac{90}{226}i = -\frac{3}{113} - \frac{45}{113}i$$

35) The answer is on the following graph.

First, simplify the inequality: $2y > 4x^2 \rightarrow \frac{2y}{2} > \frac{4x^2}{2} \rightarrow y > 2x^2$

Now, graph the quadratic $y = 2x^2$

Plug in some values for x and solve for y.

$x = 0 \rightarrow y = 2(0)^2 = 0, x = 1 \rightarrow y = 2(1)^2 = 2, x = -1 \rightarrow y = 2(-1)^2 = 2$

$x = 2 \rightarrow y = 2(2)^2 = 4, \;\; x = -2 \rightarrow y = 2(-2)^2 = 4$

Since the inequality sign is greater than (>), we need to use dash lines.

Now, choose a testing point inside the parabola. Let's choose $(0, 2)$.

$y > 2x^2 \rightarrow 2 > 2(0)^2 \rightarrow 2 > 0$

This is true. So, inside the parabola is the solution section.

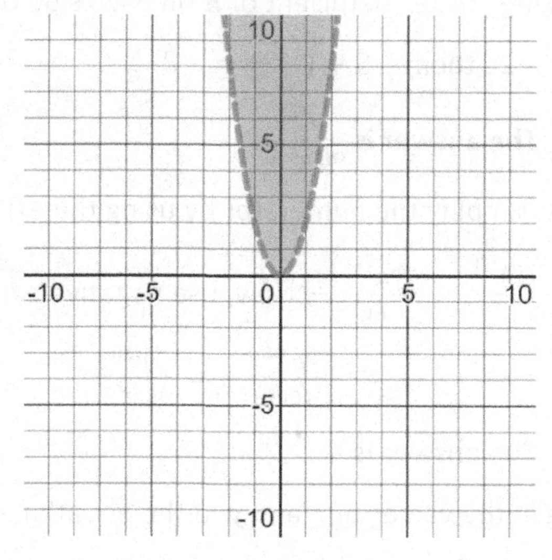

ALEKS Mathematics Practice Test 9

1) The answer is $4x^5 - 7x + 1$.

Factor the expression: $12x^8 - 21x^4 + 3x^2 = 3x^3(4x^5 - 7x + 1)$.

Therefore, the correct list of factors is $4x^5 - 7x + 1$.

2) The answer is $x = \frac{\log 11 - \log 3}{2 \log 7}$.

Use the logarithm rule that if $f(x) = g(x)$, then: $\log_a f(x) = \log_a g(x)$.

Therefore: $3(7^{2x}) = 11 \rightarrow \log 3(7^{2x}) = \log 11$.

Next: $\log_a(x \cdot y) = \log_a x + \log_a y \rightarrow \log 3(7^{2x}) = \log 3 + \log 7^{2x}$, and $\log_a x^b = b \log_a x \rightarrow \log 7^{2x} = 2x \log 7$. So, $3(7^{2x}) = 11 \rightarrow \log 3 + 2x \log 7 = \log 11$.

Rewrite this as $2x \log 7 = \log 11 - \log 3 \rightarrow x = \frac{\log 11 - \log 3}{2 \log 7}$.

3) The answers are 13 *and* 15.

Let's put x for a smaller integer. Then, the two integers are x and $x + 2$. (The difference of any two consecutive odd (or even) integers is 2). The sum of two integers is 28. Write the equation and solve for x:

$$x + (x + 2) = 28 \rightarrow 2x + 2 = 28 \rightarrow 2x = 26 \rightarrow \frac{2x}{2} = \frac{26}{2} \rightarrow x = 13$$

The smaller integer is 13 and the bigger integer is 15 $(13 + 2 = 15)$.

4) The answers are: Slope is -4 and y-intercept is -2.

Write the equation in slope-intercept form. The slope-intercept form of the equation of a line is $y = mx + b$. Then: $-8x - 2y = 4 \rightarrow -8x - 2y + 8x = 4 + 8x \rightarrow$

$$-2y = 8x + 4 \rightarrow \frac{-2y}{-2} = \frac{8x}{-2} + \frac{4}{-2} \rightarrow y = -4x - 2 \rightarrow$$

The slope-intercept form of the line is: $y = -4x - 2$.

Then, the slope is -4 and the y $-$intercept is -2.

Now, you can graph the line.

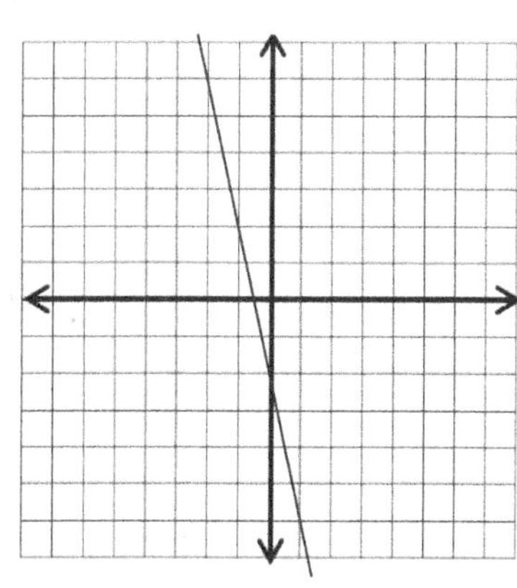

5) The answer is $\left(-\frac{2}{3}, \frac{2}{3}\right)$.

To find the vertex of the quadratic function $g(x) = 3x^2 + 4x + 2$, we need to use the formula $-\frac{b}{2a}$ to find the x –coordinate of the vertex, and then plug that value into the equation to find the corresponding y –coordinate.

In this case, $a = 3$ and $b = 4$, so the x –coordinate of the vertex is given by:

x –coordinate of vertex: $-\frac{b}{2a} = -\frac{4}{2(3)} = -\frac{4}{6} = -\frac{2}{3}$.

To find the y –coordinate of the vertex, we can plug in $x = -\frac{2}{3}$ into the equation for $g(x)$ and simplify:

y –coordinate of vertex: $g\left(-\frac{2}{3}\right) = 3\left(-\frac{2}{3}\right)^2 + 4\left(-\frac{2}{3}\right) + 2 = \frac{2}{3}$.

Therefore, the vertex of the graph of the quadratic function $g(x) = 3x^2 + 4x + 2$ is $\left(-\frac{2}{3}, \frac{2}{3}\right)$.

6) The answer is 3.

Solving systems of equations by elimination: multiply the first equation by (-2), then add it to the second equation.

$$\begin{array}{c} -2(2x + 5y = 11) \\ \underline{4x - 2y = -14} \end{array} \Rightarrow \begin{array}{c} -4x - 10y = -22 \\ 4x - 2y = -14 \end{array} \Rightarrow -12y = -36 \Rightarrow y = 3$$

7) The answer is $x_n = \left(-\frac{1}{2}\right)^{n-3}$.

Considering the sequence, the starting term is 4. To find the common ratio, divide a_2 by a_1: $r = \frac{a_2}{a_1} = \frac{-2}{4} = -\frac{1}{2}$. Now, use the geometric sequence formula: $x_n = a_1 r^{(n-1)}$. Then:

$$x_n = 4\left(-\frac{1}{2}\right)^{(n-1)} \rightarrow x_n = 2^2\left(-\frac{1}{2}\right)^{(n-1)} = (-2)^2\left(-\frac{1}{2}\right)^{(n-1)} = \left(-\frac{1}{2}\right)^{-2}\left(-\frac{1}{2}\right)^{(n-1)} = \left(-\frac{1}{2}\right)^{(n-1-2)} \rightarrow x_n = \left(-\frac{1}{2}\right)^{n-3}.$$

8) The answer is $-\frac{55a-12y}{12a}$.

To write this expression as a single fraction, we need to find a common denominator.

The common denominator of $12a$ and $8a$ is $24a$. Then:

$$-4 + \frac{2a - 6y}{12a} - \frac{6a + 4y}{8a} = \frac{-4(24a)}{24a} + \frac{2(2a - 6y)}{24a} - \frac{3(6a + 4y)}{24a}$$

Now, simplify the numerators and combine:

$$\frac{-4(24a)}{24a}+\frac{2(2a-6y)}{24a}-\frac{3(6a+4y)}{24a}=\frac{-96a}{24a}+\frac{4a-12y}{24a}-\frac{18a+12y}{24a}=$$

$$\frac{-96a+4a-12y-18a-12y}{24a}=\frac{-110a-24y}{24a}$$

Divide both numerator and denominator by 2. Then:

$$\frac{-110a-24y}{24a}=\frac{-12y-55a}{12a}$$

9) The answer is -8.

$g(x)=-2$, then $f(g(x))=f(-2)=3(-2)^3+5(-2)^2+2(-2)=-24+20-4=-8$

10) The answer is $25,000<(4,334-2,712)t$.

The savings each year from installing the geothermal heating system will be the average annual energy cost for the home before the geothermal heating system installation minus the average annual energy cost after the geothermal heating system installation, which is $(4{,}334-2{,}712)$ dollars. In t years, the savings will be $(4{,}334-2{,}712)t$ dollars. Therefore, the inequality that can be solved to find the number of years after installation at which the total amount of energy cost savings will exceed (be greater than) the installation cost, \$25,000, is $25{,}000<(4{,}334-2{,}712)t$.

11) The answer is $\frac{1}{2}$.

The relationship among all sides of right triangle $30°-60°-90°$ is provided in the following triangle: Sine of $30°$ equals to: $\frac{opposite}{hypotenuse}=\frac{x}{2x}=\frac{1}{2}$

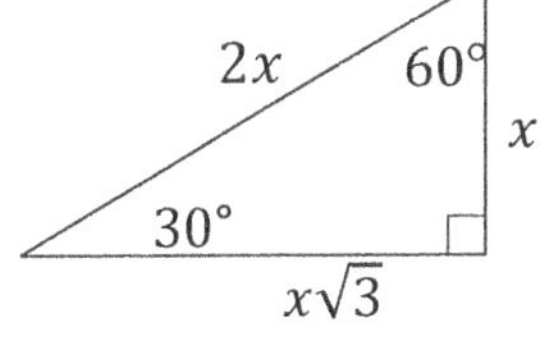

12) The answer is $155°$.

The angles $\boldsymbol{x}$ and $\mathbf{25}$ are supplementary angles. Therefore: $\boldsymbol{x}+\mathbf{25°}=\mathbf{180°}\rightarrow$

$\boldsymbol{x}=\mathbf{180°}-\mathbf{25°}=\mathbf{155°}$

13) The answer is $30°$.

The sum of supplement angles is 180. Let x be that angle. Therefore, $x+5x=180$

$6x=180$, divide both sides by 6: $x=30$ degrees.

14) The answer is $c = 0.35(60h)$.

$0.35 per minute to use the car. This per-minute rate can be converted to the hourly rate using the conversion 1 hour = 60 minutes, as shown below.

$$\frac{\$0.35}{minute} \times \frac{60\ minutes}{1\ hours} = \frac{\$(0.35 \times 60)}{hour}$$

Thus, the car costs $(0.35 × 60) per hour.

Therefore, the cost c, in dollars, for h hours of use is $c = (0.35 \times 60)h$,

Which is equivalent to $c = 0.35(60h)$

15) The answer is 38.

Let x be the smallest number. Then, these are the numbers: $x, x+1, x+2, x+3$,

and $x+4$. $average = \frac{sum\ of\ terms}{number\ of\ terms} \Rightarrow 40 = \frac{x+(x+1)+(x+2)+(x+3)+(x+4)}{5} \Rightarrow 40 = \frac{5x+10}{5} \Rightarrow$

$200 = 5x + 10 \Rightarrow 190 = 5x \Rightarrow x = 38$

The smallest number is **38**.

16) The answer is $\frac{\sqrt{15}}{4}$.

$\sin A = \frac{1}{4} \Rightarrow$ Since $\sin\theta = \frac{opposite}{hypotenuse}$, we have the following right triangle. Then:

$c = \sqrt{4^2 - 1^2} = \sqrt{16-1} = \sqrt{15}$

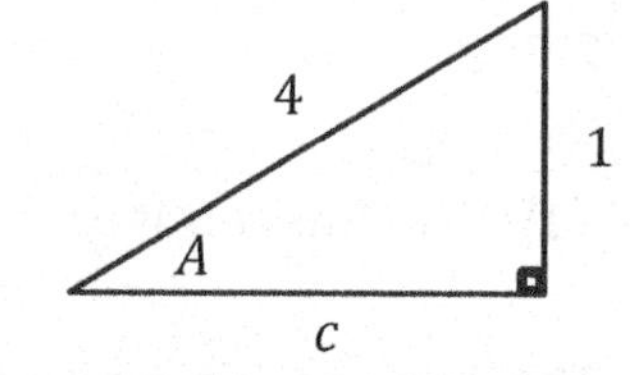

$$\cos A = \frac{adjacent}{hypotenuse} = \frac{\sqrt{15}}{4}$$

17) The answer is $16\,\pi$.

The equation of a circle in standard form is $(x-h)^2 + (y-k)^2 = r^2$, where r is the radius of the circle. In this circle the radius is 4. $r^2 = 16 \rightarrow r = 4$, $(x+2)^2 + (y-4)^2 = 16$, Area of a circle: $A = \pi r^2 = \pi\,(4)^2 = 16\,\pi$.

18) The answer is $-\frac{1}{2}$.

The equation of a line in slope-intercept form is: $y = mx + b$, solve for y. $6x - 3y = 15 \Rightarrow -3y = 15 - 6x \Rightarrow y = \frac{(15-6x)}{-3} \Rightarrow y = 2x - 5$, the slope of the line is 2. The slope of the line perpendicular to this line is: $m_1 \times m_2 = -1 \Rightarrow 2 \times m_2 = -1 \Rightarrow m_2 = -\frac{1}{2}$.

19) The answer is $x = \frac{ln(10)-2-ln\,3}{2\,ln\,5-ln\,3}$.

Use the logarithm rule that if $f(x) = g(x)$, then: $log_a f(x) = log_a g(x)$.

Therefore: $e^2 3^{1-x} = \frac{10}{5^{2x}} \rightarrow ln(e^2 3^{1-x}) = ln\left(\frac{10}{5^{2x}}\right)$.

Next: $log_a(x \cdot y) = log_a x + log_a y \rightarrow ln(e^2 3^{1-x}) = ln(e^2) + ln(3^{1-x})$,

$log_a\left(\frac{x}{y}\right) = log_a x - log_a y \rightarrow ln\left(\frac{10}{5^{2x}}\right) = ln(10) - ln(5^{2x})$,

Then: $ln(e^2 3^{1-x}) = ln\left(\frac{10}{5^{2x}}\right) \rightarrow ln(e^2) + ln(3^{1-x}) = ln(10) - ln(5^{2x})$.

And $log_a x^b = b\,log_a x \rightarrow ln(e^2) = 2\,ln\,e$, $ln(3^{1-x}) = (1-x)\,ln\,3$, and $ln(5^{2x}) = (2x)\,ln\,5$.

So: $e^2 3^{1-x} = \frac{10}{5^{2x}} \rightarrow 2\,ln\,e + (1-x)\,ln\,3 = ln(10) - (2x)\,ln\,5$.

Use the rule: $log_a a = 1$. Simplify:

$2 + ln\,3 - x\,ln\,3 = ln(10) - (2x)\,ln\,5$

Rewrite this as: $(2x)\,ln\,5 - x\,ln\,3 = ln(10) - ln\,3 - 2 \rightarrow x(2\,ln\,5 - ln\,3) = ln(10) - 2 - ln\,3$

$\rightarrow x = \frac{ln(10)-2-ln\,3}{2\,ln\,5-ln\,3}$.

20) The answer is 42.5.

First, find the number. Let x be the number. Write the equation and solve for x. $\mathbf{140\%}$ of a number is $\mathbf{70}$, then: $\mathbf{1.4 \times x = 70 \Rightarrow x = 70 \div 1.4 = 50}$, $\mathbf{85\%}$ of $\mathbf{50}$ is: $\mathbf{0.9 \times 50 = 42.5}$

21) The answer is 1.

The cotangent is the reciprocal of tangent: $tangent\ \beta = \frac{1}{cotangent\ \beta} = \frac{1}{1} = 1$

22) The answer is $\frac{3}{4}x$.

Find the common denominator and simplify the expression.

$$\sqrt{\frac{x^2}{2} + \frac{x^2}{16}} = \sqrt{\frac{8x^2}{16} + \frac{x^2}{16}} = \sqrt{\frac{9x^2}{16}} = \sqrt{\frac{9}{16}x^2} = \sqrt{\frac{9}{16}} \times \sqrt{x^2} = \frac{3}{4} \times x = \frac{3}{4}x$$

23) The answer is $\frac{8}{17}$.

$tan\,\theta = \frac{opposite}{adjacent}$, and $tan\,x = \frac{8}{15}$, therefore, the opposite side of the angle x is 8 and the adjacent side is 15. Let's draw the triangle.

Using the Pythagorean theorem, we have: $a^2 + b^2 = c^2 \rightarrow 8^2 + 15^2 = c^2 \rightarrow 64 + 225 = c^2 \rightarrow$

$c = 17, sin\ x = \frac{opposite}{hypotenuse} = \frac{8}{17}$

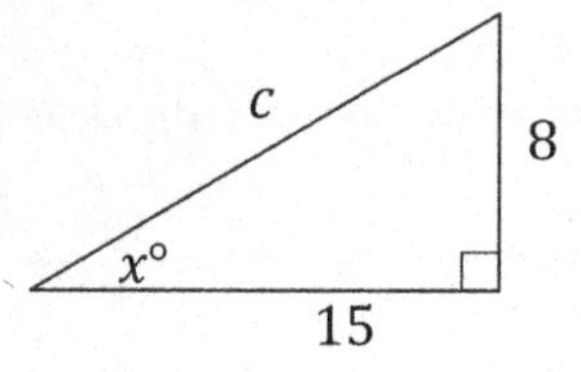

24) The answers are $-4 \leq y \leq 3$.

The possible y values are between -4 and 3. Range: $-4 \leq y \leq 3$

25) The answer is $0, -2, -4$.

First, factor the function: $f(x) = x^3 + 6x^2 + 8x = x\,(x + 4)(x + 2)$, to find the zeros, $f(x)$ should be zero. $f(x) = x\,(x + 4)(x + 2) = 0$, therefore, the zeros are: $x = 0$,

$(x + 4) = 0 \Rightarrow x = -4$, $(x + 2) = 0 \Rightarrow x = -2$

26) The answer is $\frac{11x-3}{x^4}$.

First, find a common denominator for both fractions in the expression $\frac{6}{x^3} + \frac{5x-3}{x^4}$.

The common denominator is x^4. Now, we can combine like terms into a single numerator over the denominator:

$$\frac{6x}{x^3} + \frac{5x - 3}{x^4} = \frac{6x + (5x - 3)}{x^4} = \frac{11x - 3}{x^4}$$

27) The answer is 25.

The angles on a straight line add up to 180 degrees. Then: $x + 25 + y + 2x + y = 180$

$\rightarrow 3x + 2y = 180 - 25, x = 35 \rightarrow 3(35) + 2y = 155 \rightarrow 2y = 155 - 105 = 50 \rightarrow y = 25$

28) The answer is $\{4, 8, 12\}$.

$B \cup C$ represents the union of B and C. Then $B \cup C = \{1,3,4,7,8,11,12,16,18\}$. $(B \cup C) \cap A$ represents $B \cup C$ intersection A. It means that $(B \cup C) \cap A = \{4,8,12\}$.

29) The answer is $28\ in$.

Let x be the width of the rectangle. Use Pythagorean theorem: $a^2 + b^2 = c^2$

$x^2 + 8^2 = 10^2 \Rightarrow x^2 + 64 = 100 \Rightarrow x^2 = 100 - 64 = 36 \Rightarrow x = 6$

The width of the rectangle is **6** inches. Then:

The perimeter of the rectangle = $2\,(length + width) = 2\,(8 + 6) = 2\,(14) = 28\ in$

30) The answer is on the graph.

To graph this line, we need to find two points. When x is zero the value of y is -5. And when y is 0 the value of x is 2.5.

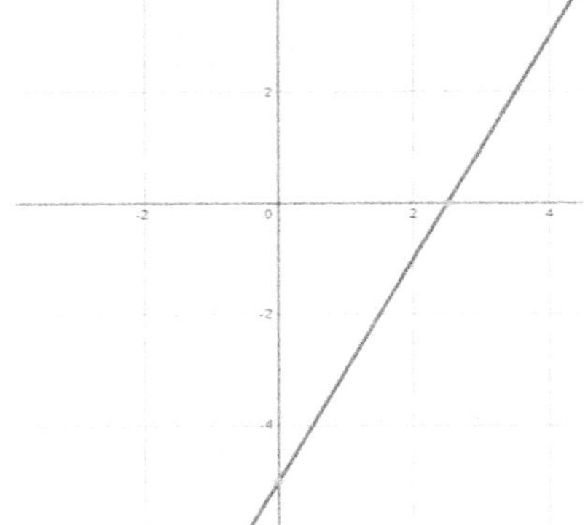

$x = 0 \rightarrow y + 3 = 2(0 - 1) \rightarrow y = -5$

$y = 0 \rightarrow 0 + 3 = 2(x - 1) \rightarrow x = 2.5$

Now, we have two points:

$(0, -5)$ and $(2.5, 0)$.

Find the points on the coordinate plane and graph the line. Remember that the slope of the line is 2.

31) The answer is $224 + 62i$.

Use the FOIL (First, Out, In, Last) method to multiply two imaginary expressions:

$(-8)(-6) + (-8)(-16i) + (11i)(-6) + (11i)(-16i) = 48 + 128i - 66i - 176i^2$

Combine like terms $(+128i - 66i)$ and simplify:

$48 + 128i - 66i - 176i^2 = 48 + 62i - 176i^2$

$i^2 = -1$, then: $48 + 62i - 176i^2 = 48 + 62i - 176(-1) = 48 + 62i + 176 = 224 + 62i$

32) The answer is 4.

Using $(x - y)^3 = x^3 - y^3 - 3xy(x - y)$, so:

$(a - b)^3 = a^3 - b^3 - 3ab(a - b)$.

Substitute $a - b = 2$ and $a^3 - b^3 = 32$. Therefore,

$(2)^3 = 32 - 3ab(2) \rightarrow 8 = 32 - 6ab \rightarrow 6ab = 24 \rightarrow ab = 4$.

33) The answer is $2(x - 6)(x + 6)$ or $2x^2 - 72$.

Find the factors of the denominators: $\frac{3x}{x^2-36} = \frac{3x}{(x-6)(x+6)}$ and $\frac{4}{2x-12} = \frac{4}{2(x-6)}$

Since the factor $(x - 6)$ is common in both denominators, then, the least common

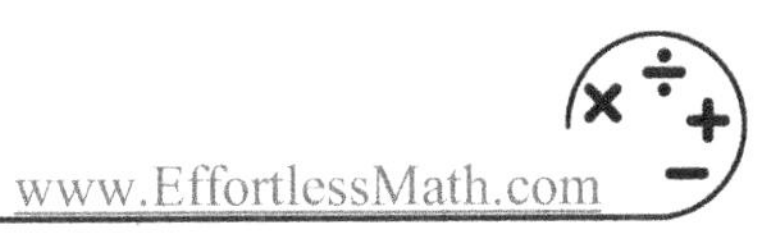

denominator is: $2(x-6)(x+6) = 2x^2 - 72$

34) The answer is 3.9.

To find AC, use $sine\ B$. Then:

$sine\ \theta = \frac{opposite}{hypotenuse}$. $sine\ 40° = \frac{AC}{6} \rightarrow 6 \times sine\ 40° = AC$,

Now use a calculator to find $sine\ 40°$. $sine\ 40° \approx\ 0.643 \rightarrow AC\ \approx 3.86$

35) The answer is -1.

To find the value of $\boldsymbol{f(8)}$, make $\boldsymbol{x^3 = 8}$. Then, $\boldsymbol{x^3 = 8 \rightarrow x = 2}$. Finally,

$\boldsymbol{x = 2 \rightarrow f(2^3) = 2(2) - 5 = -1 \rightarrow f(8) = -1}$.

ALEKS Mathematics Practice Test 10

1) The answer is $y = |x + 2| - 3$.

The general form of an absolute function is: $f(x) = a|x - h| + k$.

Since the graph opens upward with a slope of 1, then a is positive one. The graph has moved 3 units down, so the value of k is -3, it has and moved 2 units left, so the value of h is -2. Then, $y = |x + 2| - 3$.

2) The answer is $168x^5u^7$.

Apply exponent rules: $x^a \times x^b = x^{a+b}$

Then: $4x^4u^5.6u^2.7x = 168x^5u^7$

3) The answer is 2.

Considering the factorial, we have: $(n + 2)! = (n + 2)(n + 1)n(n - 1)!$. So, substitute:

$\frac{(n+2)!}{(n-1)!} = 24 \to \frac{(n+2)(n+1)n(n-1)!}{(n-1)!} = 24 \to (n + 2)(n + 1)n = 24.$

The answer to the problem is the answer to this question: Which product of three consecutive natural numbers is 24? The result is clearly obtained for $x = 2$.

4) The answer is $(2x - 5)(2x + 3)$.

Find the factors of the product of a and c (ac) that add to b. $ac = 4(-15) = -60$, $b = -4$

Write bx as a sum or difference using the factors from step 1.$-4x = -10x + 6x$

Divide the polynomial into 2 groups. $4x^2 - 10$ and $+6x - 15$.

Take the common factor out of both groups. Then: $2x(2x - 5) + 3(2x - 5)$

Factor out the common binomial factor. $2x(2x - 5) + 3(2x - 5) = (2x - 5)(2x + 3)$

5) The answer is 25.

Let x be the number. Write the equation and solve for x. $\frac{2}{3} \times 15 = \frac{2}{5}.x \Rightarrow \frac{2\times15}{3} = \frac{2x}{5}$, use cross multiplication to solve for x. $5 \times 30 = 2x \times 3 \Rightarrow 150 = 6x \Rightarrow x = 25$

6) The answer is 0.

If the value of $|x-3|+3$ is equal to 0, then $|x-3|+3=0$. Subtracting 3 from both sides of this equation gives $|x-3|=-3$. The expression $|x \; - \; 3|$ on the left side of the equation is the absolute value of $x \; - \; 3$, and the absolute value can never be a negative number.

Thus $|x-3|=-3$ has no solution. Therefore, there are no values for x for which the value of $|x-3|+3$ is equal to 0.

7) The answer is -38.

Use PEMDAS (order of operation):

$$5+8\times(-3)-\;[4+22\times5]\div6=5+8\times(-3)-[4+110]\div$$
$$=5+8\times(-3)-\;[114]\div6=5+(-24)-19=5+(-24)-19=5-43$$
$$=-38$$

8) The answer is 10.

If the score of Mia was 40, then the score of Ava is 20. Since the score of Emma was half that of Ava, therefore, the score of Emma is 10.

9) The answer is $-\sqrt{3}$.

$$tan\ \frac{2\pi}{3}=\frac{sin\frac{2\pi}{3}}{cos\frac{2\pi}{3}}=\frac{\frac{\sqrt{3}}{2}}{-\frac{1}{2}}=-\sqrt{3}$$

10) The answer is $x=15$.

To solve for x, isolate the radical on one side of the equation.

Divide both sides by 4. Then: $4\sqrt{2x+6}=24\rightarrow\frac{4\sqrt{2x+6}}{4}=\frac{24}{4}\rightarrow\sqrt{2x+6}=6$

Square both sides: $\left(\sqrt{(2x+6)}\right)^2=6^2$, Then: $2x+6=36\rightarrow2x=30\rightarrow x=15$

Substitute x by 15 in the original equation and check the answer:

$x=15\rightarrow4\sqrt{2(15)+6}=4\sqrt{36}=4(6)=24$

11) The answer is $-\frac{2}{3}\boldsymbol{log\,y}-\boldsymbol{2\,log\,x}-\boldsymbol{1}$.

Use the logarithm rule: $log_a x-log_a y=log_a\frac{x}{y}$, we get: $log\frac{\sqrt[3]{y}}{10yx^2}=log\sqrt[3]{y}-log\,10yx^2$.

Using the logarithm rule: $log_a(x\cdot y)=log_a x+log_a y$, we get:

$log\ 10yx^2 = log\ 10 + log\ y + log\ x^2$.

We know that $\sqrt[3]{y} = y^{\frac{1}{3}}$. By substituting, we get: $log\frac{\sqrt[3]{y}}{10yx^2} = log\ y^{\frac{1}{3}} - (log\ 10 + log\ y + log\ x^2)$.

Simplify:

$log\frac{\sqrt[3]{y}}{10yx^2} = log\ y^{\frac{1}{3}} - log\ 10 - log\ y - log\ x^2$.

Now, use the logarithm rules: $log_a\ x^n = n\ log_a\ x$ and $log_a\ a = 1$. Then: $log\ y^{\frac{1}{3}} = \frac{1}{3}log\ y$, $log\ x^2 = 2\ log\ x$ and $log\ 10 = 1$. Finally: $log\frac{\sqrt[3]{y}}{10yx^2} = \frac{1}{3}log\ y - log\ y - 2\ log\ x - 1 = -\frac{2}{3}log\ y - 2\ log\ x - 1$.

12) The answer is $(x + 4)^2 + (y - 2)^2 = 1$.

First, find the radius of the circle. The circumference of a circle $= 2\pi \Rightarrow$

circumference $= 2\pi r = 2\ \pi \Rightarrow r = 1$

The equation of a circle in the coordinate plane: $(x - h)^2 + (y - k)^2 = r^2 \Rightarrow$

Center: (h, k) and radius: r, center: $(-4, 2) \Rightarrow h = -4, k = 2$

Then, the equation of the circle is: $(x + 4)^2 + (y - 2)^2 = 1$

13) The answer is $x = 31$.

The two angles are complementary angles (their sum is 90 degrees). Then:

$2x - 6 + (x + 3) = 90 \rightarrow 3x - 3 = 90 \rightarrow 3x = 93 \rightarrow x = 31$

14) The answer is $x = 7$.

Let x be the number. Write the equation and solve for x. $(28 - x) \div x = 3$

Multiply both sides by x. $(28 - x) = 3x$, then add x on both sides. $28 = 4x$, now divide both sides by 4. $x = 7$

15) The answers are $475°$ and $-245°$.

Coterminal angles are equal angles. To find a Coterminal of an angle, add or subtract 360 degrees (or $2\ \pi$ for radians) to the given angle. Then:

$115^\circ + 360^\circ = 475^\circ$

$115^\circ - 360^\circ = -245^\circ$

16) The answer is $x \geq 2$.

To solve the inequality, bring the variable x to one side by adding or subtracting. Then: $8x - 4 \geq -2x + 16 \rightarrow 8x + 2x - 4 \geq -2x + 16 + 2x \rightarrow 10x - 4 + 4 \geq 16 + 4$

$\rightarrow 10x \geq 20 \rightarrow x \geq \frac{20}{10} \rightarrow x \geq 2$

Since the variable is greater than or equal to 2, then we need to find 2 on the number line and draw an closed circle on it. Then, draw an arrow to the right.

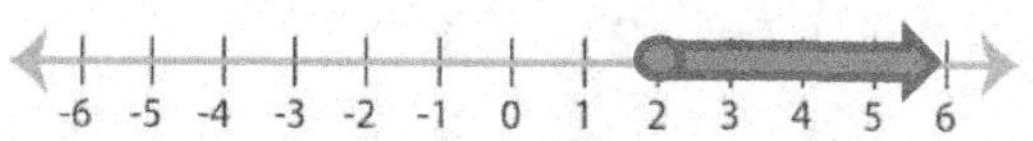

17) The answer is $h(x) = x^2(x-1)(x^2+x+1)$.

Factor the common expression. So, $x^5 - x^2 = x^2(x^3 - 1)$. Now, use the polynomial identity formula as $x^3 - y^3 = (x-y)(x^2 + xy + y^2)$, then: $x^3 - 1 = (x-1)(x^2+x+1)$. Here, $x^2 + x + 1$ is a quadratic equation. Evaluate Δ. Then,

$\Delta = (1)^2 - 4(1)(1) = 1 - 4 = -3 < 0$.

Accordingly, the equation $x^2 + x + 1$ has no other factor. Therefore, we have:

$h(x) = x^2(x-1)(x^2+x+1)$.

18) The answer is $x = -\frac{8}{19}$.

Use distributive property to simplify $3(12x + 4)$ and $-2(x + 2)$. Then:

$3(12x + 4) = 36x + 12$ and $-2(x + 2) = -2x - 4$

Isolate the variable: $36x + 12 = -2x - 4$, subtract 12 from both sides.

$\rightarrow 36x + 12 - 12 = -2x - 4 - 12 \rightarrow 36x = -2x - 16$, add $2x$ to both sides:

$\rightarrow 36x + 2x = -2x - 16 + 2x \rightarrow 38x = -16$. Divide both sides by 38:

$\rightarrow \frac{38x}{38} = \frac{-16}{38} \rightarrow x = -\frac{8}{19}$

19) The answer is 5.

Use distance of two points formula: $d = \sqrt{(x_A - x_B)^2 + (y - y_B)^2} \rightarrow$

$d = \sqrt{(1-(-2))^2 + (3-7)^2} = \sqrt{(3)^2 + (-4)^2} = \sqrt{9+16} = \sqrt{25} = 5$

The distance between the two points on the coordinate plane is 5 units.

20) The answer is: Solution $x < -2$ or $x > 8$, Interval Notation: $(-\infty, -2) \cup (8, \infty)$.

Factor the numerator: $\frac{6x+12}{x-8} > 0 \rightarrow \frac{6(x+2)}{x-8} > 0$

Divide both sides by 6: $\rightarrow \frac{\frac{6(x+2)}{x-8}}{6} > \frac{0}{6} \rightarrow \frac{x+2}{x+8} > 0 \rightarrow$ Find the signs of the factors $\frac{x+2}{x+8}$.

Plug in some values of x and check the solutions. Only $x < -2$ or $x > 8$ work in the inequality.

The interval notation: $(-\infty, -2) \cup (8, \infty)$

21) The answer is 31.

Find the sum of five numbers.

$average = \frac{sum\ of\ terms}{number\ of\ terms} \Rightarrow 26 = \frac{sum\ of\ 5\ numbers}{5} \Rightarrow sum\ of\ 5\ numbers = 26 \times 5 = 130$

The sum of 5 numbers is 130. If a sixth number 56 is added, then the sum of 6 numbers is $130 + 56 = 186$. The new average is: $\frac{sum\ of\ 6\ numbers}{6} = \frac{186}{6} = 31$

22) The answer is -7.

We know that if $x + y + z = 0$, then $x^3 + y^3 + z^3 = 3xyz$. Since $a + 2b = c$, so, $a + 2b - c = 0 \rightarrow a + 2b + (-c) = 0$. Therefore, $a^3 + 8b^3 - c^3 = 3a(2b)(-c) = -6abc$.

Finally, $a^3 + 8b^3 - c^3 = -6 \times 12 = -72$.

23) The answer is $4a^2 + b^2 - 4ab - 2b + 4a + 1 - 4ca^2 - cb^2 + 4cab + 2cb - 4ca - c$.

Considering that $(x + y + z)^2 = x^2 + y^2 + z^2 + 2xy + 2yz + 2zx$, then:

$(2a - b + 1)^2 = (2a)^2 + (-b)^2 + 1^2 + 2(2a)(-b) + 2(-b)1 + 2(2a)1$.

Simplified, this would be:

$(2a - b + c)^2 = 4a^2 + b^2 - 4ab - 2b + 4a + 1$.

Now, multiply the above by $(1-c)$:

$(1-c)(4a^2+b^2-4ab-2b+4a+1)=$

$4a^2+b^2-4ab-2b+4a+1-4ca^2-cb^2+4cab+2cb-4ca-c.$

24) The answer is 39.

Let x be the integer. Then: $2x-5=73$, Add 5 to both sides: $2x=78$, Divide both sides by 2: $x=39$.

25) The answer is $(-1+\sqrt{6})$ and $(-1-\sqrt{6})$.

Use quadratic formula: $x_{1,2}=\frac{-b\pm\sqrt{b^2-4ac}}{2a}$

The quadratic equation in standard form is: $ax^2+bx+c=0$

For the equation: $x^2+2x-5=0 \Rightarrow$ Then: $a=1$, $b=2$ and $c=-5$

$x=\frac{-2+\sqrt{2^2-4\times 1\times(-5)}}{2\times 1}=\frac{-2+\sqrt{4-(-20)}}{2}=\frac{-2+\sqrt{24}}{2}=\frac{-2+\sqrt{6\times 4}}{2}=\frac{-2+2\sqrt{6}}{2}=\frac{2(-1+\sqrt{6})}{2}=-1+\sqrt{6}$ or $x=\frac{-2-\sqrt{2^2-4\times 1\times(-5)}}{2\times 1}=-1-\sqrt{6}$

26) The answer is on the graph.

To draw the graph of $y\le -2x-2$, you first need to graph the line: $y=-2x-2$

Since there is a less than ($\le$) sign, draw a solid line.

The slope is -2 and y-intercept is -2.

Then, choose a testing point and substitute the value of x and y from that point into the inequality. The easiest point to test is the origin: $(0,0)$

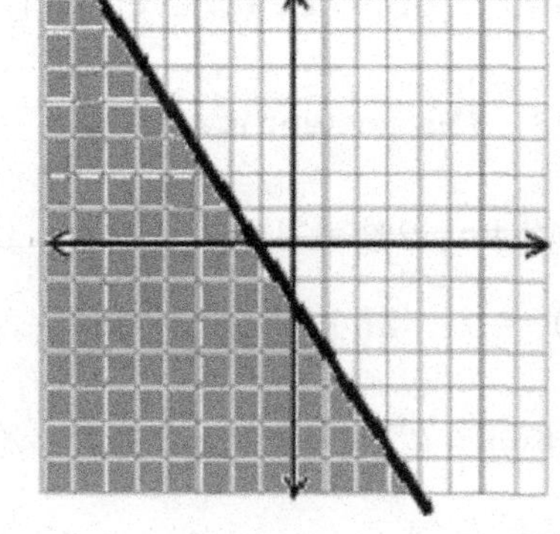

$(0,0)\rightarrow y\le -2x-2\rightarrow 0\le -2(0)-2\rightarrow 0\le -2$

This is incorrect! 0 is not less than -2. So, the left side of the line is the solution of this inequality.

27) The answer is $\{x: x\ne 1,3\}$.

First, simplify the expression $\frac{x^3-1}{x^2-4x+3}$. For this purpose, write the numerator and denominator as the product of their factors. The numerator using the formula $x^3 - y^3 = (x - y)(x^2 + xy + y^2)$, becomes $x^3 - 1 = (x - 1)(x^2 + x + 1)$, and the denominator by the formula

$(x + a)(x + b) = x^2 + (a + b)x + ab$,

becomes $x^2 - 4x + 3 = (x - 1)(x - 3)$. Therefore,

$\frac{x^3-1}{x^2-4x+3} = \frac{(x-1)(x^2+x+1)}{(x-1)(x-3)} = \frac{(x^2+x+1)}{(x-3)}$.

Now, find the zeros in the denominator. Then remove them from the set of real numbers.

$x - 3 = 0 \rightarrow x = 3$ and $x - 1 = 0 \rightarrow x = 1$

Therefore, the domain of the function f is all real numbers except 3 and 1.

28) The answer is 32.

Logarithm is another way of writing exponent. $log_b{}^y = x$ is equivalent to $y = b^x$.

Rewrite the logarithm in exponent form: $log_2 x = 5 \rightarrow 2^5 = x \rightarrow x = 32$

29) The answer is 2.

To solve this problem first solve the equation for c. $\frac{c}{b} = 5$

Multiply by b on both sides. Then: $b \times \frac{c}{b} = 5 \times b \rightarrow c = 5b$. Now to calculate $\frac{10b}{c}$, substitute the value for c into the denominator and simplify. $\frac{10b}{c} = \frac{10b}{5b} = \frac{10}{5} = 2$

30) The answer is 60.

Plug in 140 for F and then solve for C.

$$C = \frac{5}{9}(F - 32) \Rightarrow C = \frac{5}{9}(140 - 32) \Rightarrow C = \frac{5}{9}(108) = 60$$

31) The answers are 1, 31, 465, 4, 495 and 31, 465.

Using the Pascal's triangle properties: ${}_nC_{k-1} + {}_nC_k = {}_{n+1}C_k$. So, each entry in row $n + 1$ is obtained from the sum of the previous two entries in row n. Therefore:

${}_{31}C_0 = \frac{31!}{0!(31-0)!} = \frac{31!}{31!} = 1$,

${}_{31}C_1 = {}_{30}C_0 + {}_{30}C_1 \rightarrow {}_{31}C_1 = 1 + 30 = 31$,

${}_{31}C_2 = {}_{30}C_1 + {}_{30}C_2 \rightarrow {}_{31}C_2 = 30 + 435 = 465$,

${}_{31}C_3 = {}_{30}C_2 + {}_{30}C_3 \rightarrow {}_{31}C_3 = 435 + 4{,}060 = 4{,}495,$

${}_{31}C_4 = {}_{30}C_3 + {}_{30}C_4 \rightarrow {}_{31}C_4 = 4{,}060 + 27{,}405 = 31{,}465.$

Finally, the first five terms in row 31 are 1, 31, 465, 4,495 and 31,465.

32) The answer is $y = 0$.

In a rational function, if the denominator has a bigger degree than the numerator, the horizontal asymptote is the x-axes or the line $y = 0$. In the function $f(x) = \frac{x+3}{x^2+1}$, the degree of numerator is 1 (x to the power of 1) and the degree of the denominator is 2 (x to the power of 2). Then, the horizontal asymptote is the line $y = 0$.

33) The answer is 44.

To find the value of x, use the cosine on the angle x: $cos\ \theta = \frac{adjacent}{hypotenuse} \rightarrow cos\ x = \frac{10}{14} = \frac{5}{7}$

Use a calculator to find inverse cosine: $cos^{-1}\left(\frac{5}{7}\right) = 44.41° \approx 44°$

Then: $x = 44$

34) The answer is $\frac{3\pi}{4}$.

Use this formula: Radians $=$ Degrees $\times \frac{\pi}{180}$

Radians $= 135 \times \frac{\pi}{180} = \frac{135\,\pi}{180} = \frac{3\,\pi}{4}$

35) The answer is 8.08×10^8.

Convert the first number to have the same power of 10.

$8.5 \times 10^8 = 85 \times 10^7$

Now, two numbers have the same power of 10. Factor 10^7 out.

$85 \times 10^7 - (4.2 \times 10^7) = (85 - 4.2) \times 10^7$

Subtract: $85 - 4.2 = 80.8$. Then: $(85 - 4.2) \times 10^7 = 80.8 \times 10^7 = 8.08 \times 10^8$

Effortless Math's ALEKS Online Center

... So Much More Online!

Effortless Math Online ALEKS Math Center offers a complete study program, including the following:

- ✓ Step-by-step instructions on how to prepare for the ALEKS Math test
- ✓ Numerous ALEKS Math worksheets to help you measure your math skills
- ✓ Complete list of ALEKS Math formulas
- ✓ Video lessons for ALEKS Math topics
- ✓ Full-length ALEKS Math practice tests
- ✓ And much more...

No Registration Required.

Visit **EffortlessMath.com/ALEKS** to find your online ALEKS Math resources.

Receive the PDF version of this book or get another FREE book!

Thank you for using our Book!

Do you LOVE this book?

Then, you can get the PDF version of this book or another book absolutely FREE!

Please email us at:

info@EffortlessMath.com

for details.

Author's Final Note

I hope you enjoyed reading this book. You've made it through the book! Great job!

First of all, thank you for purchasing this practice book. I know you could have picked any number of books to help you prepare for your ALEKS Math test, but you picked this book and for that I am extremely grateful.

It took me years to write this practice book for the ALEKS Math because I wanted to prepare a comprehensive ALEKS Math book to help test takers make the most effective use of their valuable time while preparing for the test.

After teaching and tutoring math courses for over a decade, I've gathered my personal notes and lessons to develop this practice test. It is my greatest hope that the practice tests in this book could help you prepare for your test successfully.

If you have any questions, please contact me at reza@effortlessmath.com and I will be glad to assist. Your feedback will help me to greatly improve the quality of my books in the future and make this book even better. Furthermore, I expect that I have made a few minor errors somewhere in this book. If you think this to be the case, please let me know so I can fix the issue as soon as possible.

If you enjoyed this book and found some benefit in reading this, I'd like to hear from you and hope that you could take a quick minute to post a review on the book's Amazon page. To leave your valuable feedback, please visit: amzn.to/2K9LEOC

Or scan this QR code.

I personally go over every single review, to make sure my books really are reaching out and helping students and test takers. Please help me help ALEKS Math test takers, by leaving a review!

I wish you all the best in your future success!

Reza Nazari

Math teacher and author

Made in the USA
Coppell, TX
25 February 2025